Burning Matters

Recent Titles in

Global and Comparative Ethnography
Edited by Javier Auyero

Burning Matters

*Life, Labor, and E-Waste Pyropolitics
in Ghana*

PETER C. LITTLE

OXFORD
UNIVERSITY PRESS

OXFORD
UNIVERSITY PRESS

Oxford University Press is a department of the University of Oxford. It furthers
the University's objective of excellence in research, scholarship, and education
by publishing worldwide. Oxford is a registered trade mark of Oxford University
Press in the UK and certain other countries.

Published in the United States of America by Oxford University Press
198 Madison Avenue, New York, NY 10016, United States of America.

Library of Congress Cataloging-in-Publication Data
Names: Little, Peter C., author.
Title: Burning matters : life, labor, and e-waste pyropolitics in
Ghana / Peter C. Little.
Description: New York, NY : Oxford University Press, [2022] |
Includes bibliographical references and index.
Identifiers: LCCN 2021018000 (print) | LCCN 2021018001 (ebook) |
ISBN 9780190934545 (hardback) | ISBN 9780190934552 (paperback) |
ISBN 9780190934576 (epub) | ISBN 9780190934569 | ISBN 9780190934583
Subjects: LCSH: Scrap metal industry—Employees—Health and
hygiene—Ghana—Accra. | Electronic waste—Environmental
aspects—Ghana—Accra. | Electronic waste—Government policy—Ghana. |
Political ecology—Ghana. | Environmental economics—Ghana. | Accra
(Ghana)—Environmental conditions. | Accra (Ghana)—Social conditions.
Classification: LCC HD9975.G6 L56 2021 (print) | LCC HD9975.G6 (ebook) |
DDC 363.72/809667—dc23
LC record available at https://lccn.loc.gov/2021018000
LC ebook record available at https://lccn.loc.gov/2021018001

DOI: 10.1093/oso/9780190934545.001.0001

3 5 7 9 8 6 4 2

Printed by Integrated Books International, United States of America

In loving memory of Mohammed Sadik and Barney Keep

Contents

Illustrations, Tables, and Maps

Photographs

Tables

Maps

Preface

All that matters is the power of falsehood.
—Achille Mbembe (2017, 51)

This book stems from my experience doing ethnographic fieldwork in Ghana, a West African state of roughly 31 million people that has since the early 2000s been the focus of a dark and toxic narrative centered on tainted electronic waste (e-waste) management challenges. This narrative is similar to other renderings of a rapidly urbanizing Africa struggling to manage accumulations of waste and garbage, the pervasive dirt of modernity, consumption, and human–machine relations. Depictions of these postcolonial urban spaces of the Global South are often couched in terms of unbearable struggle. Within the African context, this struggle is also commonly thought of as "the material expression of the failures of development and the chaos taking over the African continent" (Fredericks 2018, 23). A recent and agitated symbol of this point of friction—whether real, exaggerated, or imagined—is the open burning of e-waste to extract market metals. This complex toxic situation is the primary focus of this book. But, at its core, *Burning Matters* attempts to go beyond the prism of toxic crisis and digital discard encroaching on the urban African landscape. Instead, e-waste relationality is the focus. I focus on e-waste burning and e-waste recycling work in Ghana in relation to other truths—other things, experiences, processes, and interventions. In this way, *Burning Matters* deals with the complex ways in which the subject of e-waste and toxic struggle is simultaneously life-supporting and connected to social and economic survival and urban–hinterland relations. E-waste burning, it will be argued, activates certain relations and politics of health and environmental intervention, as well as exposes the embodied struggle of e-waste workers navigating fiery extraction and dispossession.

I focus on the contentious nature of these issues, relations, and politics in an area of Accra, Ghana, which is home to a large informal settlement known as Old Fadama and a vibrant scrap metal market known as Agbogbloshie, a place that has been described as one of the "most polluted places on Earth"

(Pure Earth, 2014). Agbogbloshie and Old Fadama have recently drawn the attention of slum tourists eager to learn about life in a rapidly urbanizing Africa. Even the *Lonely Planet* guide for Ghana includes a section highlighting the opportunity to visit and experience Ghana's hard-working "slum" life:

> Visit the Accra district of Old Fadama (Agbogbloshie) on a 3-hour walking tour, and learn about the communities who live and work in one of the city's most deprived, but also innovative, areas. Offering a sensitive introduction to a Ghanaian urban slum, the tour allows you to learn about the inventive ways that locals have forged a living through reclaiming discarded machinery and putting [it] to new use! . . . See locals hard at work salvaging discarded appliances and machinery. Learn how unwanted items are turned from waste into profit, to earn money for the families and help to keep landfill sites clear. From old computers, fridges and fans to broken microwaves, televisions, radios and industrial machinery, nothing escapes the eye of these resourceful, hard-working people. (Lonely Planet, n.d.)

This three-hour tour might help introduce visitors to some of the social, environmental, and economic realities of this urban African e-wasteland or even allow visitors to witness slum life struggle, but getting to know these "hard-working people" is an educational effort that calls for a different practice of observation, imagination, and care. *Burning Matters* relies on ethnography to learn about this deeper complexity, while also reckoning with the fact that clarity and clean storying and knowing are not necessarily the end results of even the best ethnographic research. Honestly, at times, I have felt that any effort to fully know and write about Agbogbloshie or even explain Ghana's e-waste debate is futile. But ethnography, at the very least, offers a method of learning that "binds in place that which may not at first glance appear to go together, that which seeks to break apart or move in opposite directions. It forms a unity, momentary or lasting, drawn from autochthonous elements. It carries an artisanal dimension and looks and is pieced together differently depending upon the builder and her/his goals and skills" (Renfrew 2018, 19). Being in Ghana to learn about African e-waste labor and life taught me that studying e-waste truths and toxics in Ghana had to be open and a process of enduring acknowledgment of complexity, uncertainty, and even representational discomfort and futility. Put more bluntly, doing e-waste ethnography in Agbogbloshie called for a certain kind of plasticity.

It needed to be flexible and attentive to the multitude of elements and processes that shape and transform the lives of those engaging with this dynamic
scrap metal market, a market with extremely toxic labor conditions that fuel
a global metal scrap economy and fast-growing informal e-waste trade sector
(International Labour Organization 2019).

Most consumers of modern electronics (e.g., laptops, cell phones, iPads)
and users of electrical equipment are unaware of the social, environmental,
and human health damages resulting from electronics production and
disposal (Little 2014; Gabrys 2011; Smith, Sonnenfeld, and Pellow 2006;
Grossman 2006). E-waste burners in Agbogbloshie are aware of the dangers
of the work they do, even if they don't name specific air pollutants or refer to
lead, cadmium, or polychlorinated biphenyl levels in their blood. My ethnographic goal for this project has been guided by an attempt to make visible and audible experiences commonly overshadowed by a toxic and smoky
optic that results in "easy narratives" (Lepawsky 2019) of e-waste in Ghana.
In this way, *Burning Matters* directs attention outward, toward the global political economy and ecology of e-waste, and looks inward, toward e-waste
livelihoods shaped by labor migrations, marginality, toxic exposure and illness experience, environmental health science, and the neoliberal lure of
techno-optimistic interventions in a postcolonial African context with no
shortage of "internal contortions and complexities" (Comaroff 1997, 165).
Burning Matters offers an ethnographic glimpse into the emerging anthropology and political ecology[1] of e-waste in the Global South. This effort
involves confronting the fact that "In a global state of precarity, we don't have
choices other than *looking for life in this ruin*" (Tsing 2015, 6, emphasis added).
Throughout this book, then, I show how life and labor in Agbogbloshie are
made complex by a tangle of forces, relations, and confrontations. The e-
waste labor and life in focus is understood in relation to supply-chain extraction, bodily distress and displacement, global health science, and optimistic
interventions responding to a "damaged planet" (Tsing et al. 2017) that is
only getting hotter, more viral, more rapidly cluttered with technocapital discard, and increasingly more e-damaged and scorched.

My hope is that *Burning Matters* provides an e-waste ethnographic perspective that tries to "untangle people from their shadow realities and

---

[1] A common theoretical approach used in various environmental social science and humanities
disciplines, political ecology, as I see it, seeks to critically investigate the complex cultural, political,
economic, and environmental relations and ruptures (both local and global) that matter to communities of struggle and vitality.

representations" (Biehl and Petryna 2013, 19) to, hopefully, produce other tangible e-waste truths that extend and go beyond the already dominant e-waste crisis narrative centered on and often reduced to toxic burning. It is a practice of both confronting and going beyond these fragile e-waste narratives. Following Anand (2019, 4), one can think of this book as an "ethnographic encounter with anthropology, an effort to grasp what this field does in the world, with an eye to what it might yet be . . . a possible anthropology, one to meet the challenge of uneasy times, one willing to set sail with its most imaginative kin." Over the course of four summers (2015–2018), I carried out ethnographic fieldwork in Ghana, spending most of my time in Agbogbloshie, Old Fadama, and Savelugu in the country's northern region. I relied on common ethnographic fieldwork methods, primarily interviews, participant observation, participatory photography, policy analysis, and the use of secondary sources. The research involved over 60 semistructured interviews in Agbogbloshie and the village of Savelugu (in Ghana's Northern Region). These interviews ranged from informal and semistructured interviews to life histories. Based on five research trips since 2015, and despite the usual and often unexpected edits to research foci along the way, a persistent goal has been to better understand how these e-waste workers and scrap metal "recyclers" experience life in Ghana's urban margins. This involved interviewing, surveying, witnessing, and hanging out with these workers as they navigate and experience environment, labor, health, intervention, and social life in a fiery and toxic postcolonial context where extreme forms of social, economic, and environmental injustices endure. My interviews were conducted in either English, pidgin English,[2] or Dagbani, with the help of a translator. I interviewed ($n = 36$) and surveyed ($n = 50$) people who worked as e-waste burners, as well as women working alongside e-waste burners selling a wide range of items, including food, water, cigarettes, frozen yogurt, and candy.

As is common in ethnographic research, some people you meet stand out and become integral elements in the research. My closest friend and key informant during my time in Agbogbloshie was Ibrahim Akarima, a head or "chief" e-waste burner who early in my research told me that to understand

---

[2] Pidgin English, a linguistic form marked by the use of English in concert with the local vernacular. In this linguistic case, English is the base language used, but words are dropped or tweaked in ways that disregard standardized English grammar rules. In an effort to accurately document responses and reveal interview data, we find it culturally appropriate to use the spoken vernacular of our informants.

what goes on here, I needed to understand what Dagomba people mean when they say *Barina bela buyim ni ka pa ni nyohi ni*, or "The damage is in the fire, but not in the smoke." As he tried to explain, this Dagbani proverb suggests that often struggle comes from where you do not expect it to come from. Struggle and hardship have a center point, even while they are connected to other struggles and hardships. The struggle for him and others in Agbogbloshie, he wanted me to know, was all about the struggle that comes from burning e-waste to extract valuable copper. The emissions of toxic smoke accumulating in his body and the bodies of others inhabiting this urban "wastelandia" (Chalfin 2017) are sources of damage and hurt, but the sources of struggle and hurt are multiple. This Dagbani proverb has always stuck with me, and it provides a metaphorical adhesive or anchor point for the multifarious e-waste ethnography and encounter that follows.

Acknowledgments

This book would have never happened without the invaluable support and assistance of numerous friends and colleagues over the years. First, my deepest thanks to friends and collaborators in Ghana. Without the guidance and open hearts of Sam Sandow, Ibrahim Akarima, Ambrahim Mohamed, Memuna Akarima, Abdrahaman Yakubu, Samatu, Abuu, Abiba, Awata, Sadik Mohammed, Alhassan, Abrazak, and countless other workers in Agbogbloshie, this whole project would never have gotten off the ground. They showed me around, shared their life experiences with me, taught me what to pay attention to, helped me understand their complex worlds, and ultimately gave me comfort and clarity when I felt out of place and confused. Also, the warm support, friendship, and enduring hospitality of Muhsin Barko, Frederica "Freddie" Addo, Efo "Francis" Xolali, Nii "Gabby" Apoma Akwoshongtse, Akwété Bortei-Doku, Muftahu Ango, Wazi Apoh, Shirazu, Erika Mamley Kisseih, Portia Adade Williams, Martin Oteng-Ababio, Edith Clarke, Lambert Faabeloun, and the genuine kindness of others whom I met during my many visits to Ghana were invaluable. During my trips with Ibrahim Akarima to Savelugu in Ghana's Northern Region, the former paramount chief of Savelugu and current overlord of Dagbon, Yoo Naa Abubakar Mahama II, welcomed me into his palace; and I thank him for his kindness, hospitality, and open heart.

A number of people played a part in developing my ideas and offering insight throughout the research process and during different stages of writing. My deepest thanks to friend and occasional co-author Grace Abena Akese and to Rafael "Rafa" Fernández-Font Pérez, Jenna Burrell, Darrell Fuhriman, Brenda Chalfin, Brittni Howard, Dagna Rams, Josh Lepawsky, and Deborah Pellow for meaningful conversations at various stages of this project. I have also benefited greatly from presenting material found in this book in a variety of venues. In this vein, I thank Amelia Moore, Damien White, Scott Frickel, J. Timmons Roberts, Damien Droney, Morgan Ames, Louise Badiane, Alice Mah, and Thom Davies for the invitations and intellectual exchanges. Thanks especially to Elizabeth Hoover and Daniel Renfrew for their friendship, continued inspiration, and suffering through earlier versions of this book.

I owe a great deal to the Wenner Gren Foundation for Anthropological Research for awarding me a post-PhD research grant, which allowed me to carry out research in the summers of 2016 and 2017. Additionally, I received two faculty research grants from Rhode Island College to carry out research in the summers of 2015 and 2018. Also, I thank my colleagues in the Department of Anthropology at Rhode Island College for their ongoing support and confidence in me.

I humbly thank Javier Auyero for believing in this book project early on. I also thank my editor James Cook and assistant editor Emily Mackenzie at Oxford University Press and the anonymous reviewers for their many helpful comments and critical feedback. Thanks to John Resseger for his assistance with images, to Travis Elisson for creating the maps, and to Lucy Sue Little for helping with the index. Several chapters draw from revised material that previously appeared in article form, and I would like to thank these publishers for their permission to reuse this material. Of course, any existing errors, glitches, oversights, or misinterpretations throughout are my own. As my Dagomba interlocutors might put it, "What you sow is what you will harvest" (*A yi biri sheli dina ka a yɛn che*).

Finally, I would like to thank Jenny Little for her constant and unflagging encouragement, love, friendship, and mental support in all phases of my research and writing. It is not always easy living with someone who is selfishly absorbed in writing a book, nor is it easy to deal with long periods of time away from family to do fieldwork. You are my rock, and I am forever grateful for your companionship and for our children, Lucy and Theo.

Introduction

From E-Waste Ashes to Ethnographic Intervention

*A mi niriba gari a mi tinsi [The one who knows people is better than the
one who knows places]*
—Dagbani proverb

*The eye is not filled with seeing, with only
seeing, but with understanding the sight.*
—Ed Roberson (2021), from the poem "Asked What Has Changed"

Ibrahim returned to the worker shelter with a smile. It had been a good day of
steady work burning wires. "Good copper now. Good day for copper burning,"
he shared. As Ibrahim stretched his legs and changed out of his dirty shirt, he
bummed a cigarette from Jacob, who was lying down across from us. Sitting
on seats made of discarded insulation foam, Ibrahim joked that Jacob was
being lazy today. He then lit his cigarette and answered a phone call from his
wife Memuna. He abruptly ended the conversation, calling to a nearby group of
workers to have them collect another bundle of copper wires. Keeping his eyes
on the workers, Ibrahim tucked his screen-cracked cell phone into his shorts
pocket; and after a few drags of his cigarette and a drink of water, he turned to
me and said "What we do here, everyone can see. You can see our smoke. We are
open to see. Here you can see how it goes. We make fire for copper. You see how
it goes? You see what we do? That be Agbogbloshie!"

~

Agbogbloshie is a familiar reference point in global studies and stories of
electronic waste (e-waste).[1] An urban scrap metal market in Accra, Ghana,

---

[1] E-waste has been defined and described in various ways. In its simplest version, e-waste consists
of discarded devices with electronic circuits, such as transistors and integrated circuits or microchips.

Burning Matters. Peter C. Little, Oxford University Press. © Oxford University Press 2022.
DOI: 10.1093/oso/9780190934545.003.0001

Agbogbloshie has attracted numerous international nongovernmental organizations (NGOs), environmental health scientists, slum tourists, engineers and makerspace enthusiasts, journalists, photographers, and social scientists engaged in the ever-expanding field of discard studies. Most visitors witness what they have been told about this complex, storied place.[2] The burning of biomass and various discards to recover valuable metals, like copper and aluminum, is an everyday activity in Agbogbloshie. Decomposing machinery of our late industrial age dominates the landscape, and smoke fills an atmosphere that smells of burning plastic, foam insulation, rubber, and coco husks. It smells like toxics, like trouble (Hsu 2020). Everywhere things are ablaze, and the local environment is contaminated with high concentrations of lead, mercury, cadmium, polyvinyl chloride, and plastics containing brominated flame retardants, presenting multiple health risks. These toxic risks have led to numerous interventions that seek to mitigate and even "eliminate" the burning of e-waste by primarily migrant laborers from northern Ghana. Over two decades, international and domestic organizations and agencies have turned to Agbogbloshie workers (and their bodies) to produce environmental health knowledge, reform e-waste management and policy, and modernize scrap metal extraction practices and economies. This is all important work, but as argued in this book, equally critical is the need to reckon with the "power of falsehood" (Mbembe 2017, 51) and to somehow repair the troubling and negative "gaze on Agbogbloshie" (Agyepong 2014). Knowing Agbogbloshie, knowing e-waste, needs care, repair, and enduring humanization.

Agbogbloshie is one of the oldest settlements in Accra, Ghana's bustling capital and administrative headquarters of the British Gold Coast between 1877 and 1957 (see Map I.1). Originally, this area of Accra was what the Ga people, a dominant cultural group in the coastal region, called *Agbloshishie*. In Ga (a Kwa language) *Agbogbloshie* translates as "under Agblo." Among Ga peoples, *Agblo* is the name of a deity of the river that runs through the railway community in Agbogbloshie into the Odaw River and flows to the Korle

---

But e-waste also includes electric devices, such as kitchen appliances, treadmills, cars, or any machine or product using electricity in some capacity.

[2] Following Gupta and Ferguson (1997, 6), understanding a place is largely about asking questions about place-making: "How are understandings of locality, community, and regions formed and lived? To answer this question, we must turn away from the commonsense idea that such things as locality and community are simply given or natural and turn toward a focus on social and political processes of place-making."

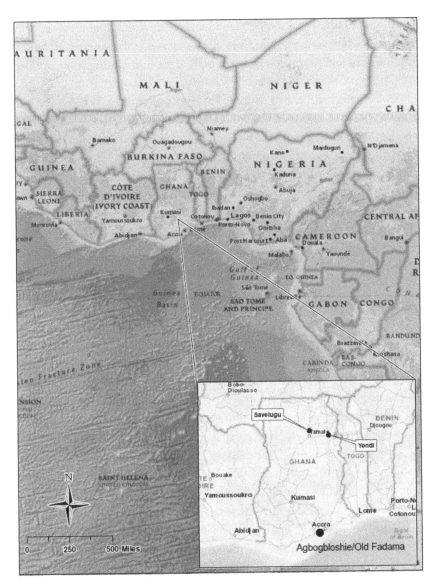

Map I.1. Ghana and Study Area.

Lagoon, which snakes its way into the Gulf of Guinea. The Agbogbloshie scrap market is about 15 acres (6.2 hectares) and rests within the Korle Lagoon, which, some international environmental NGOs claim, is one of the most polluted bodies of water in the world (Boadi and Kuitunen 2002; Brigden et al. 2008). The scrap market is also adjacent to the Agbogbloshie

Food Market, one of the largest food markets in Accra, that receives daily truckloads of consumer goods, from yams, cassava, and onions coming from Burkina Faso and Niger to luggage, bras, Q-Tips, and heating irons shipped in from China. In fact, immediately following independence from the British in 1957, Ghana's successive national leaders concentrated government and parastatal organizations in the industrial area surrounding Agbogbloshie, which eventually increased the market's importance and position in Accra, a city divided into areas of planned and unplanned growth.[3] Areas of planned growth, like the Cantonments and Airport Residential areas, are dwarfed by much larger unplanned growth areas, such as Nima, Sabon Zongo, Old Fadama, and Agbogbloshie. Places like Old Fadama are highly dense spaces where rents are much cheaper than in other parts of the city,[4] and in the adjacent Agbogbloshie market day laborers are everywhere moving commodities from place to place, from yams headed for Europe to scrap metal en route to China, India, and South Korea.

In some journalistic accounts, Agbogbloshie is referred to as "Sodom and Gomorrah," which in the Bible are depicted as two cities of sin and damnation that eventually burn to the ground. This biblical metaphor re-emerged in the 2018 documentary film *Welcome to Sodom*, directed by Christian Krönes and Florian Weigensamer, which has been rightfully criticized for reanimating problematic myths about Agbogbloshie (Velden and Oteng-Ababio 2019). As critiques of the recent film note, there is a general trend whereby Western media portrayals of Agbogbloshie miss the grounded truths and too often local complexity is dwarfed by "zombie statistics and poverty porn" (van der Velden 2019). But despite the circulation of this contentious biblical metaphor and common "hellscape" misrepresentation (Schiller 2013), workers in Agbogbloshie experience it as a complex place of work and relationships, a place of trade, market fluctuation, and vitality. They burn and build from the "ashes of the nightmare" (Kelley 2002, 196) made possible by a planetary "technocapitalism" (Suarez-Villa 2009) deeply entangled in processes of racialization, dispossession, extraction, contamination, and neoliberalization. These workers also never mention "e-waste" at all. Instead, they call attention to "metal scraps" or "copper scraps." While certainly an established urban dump and landfilled lagoon area, Agbogbloshie is a central

---

[3] Mismanagement of this rapid urban growth and inadequate planning of new settlements have led to significant challenges, including a consequent lack of sanitation facilities (Chalfin 2014).

[4] See Grant (2006) and Pellow (2008) for in-depth studies of urban residential patterns and rents in Accra.

scrap metal market where workers are drawn not to e-waste but to valuable copper scrap. Since 2010, copper scrap exports in Ghana have boomed. For example, in 2019, Ghana exported approximately 3.57 million kilograms worth $7.62 million, compared to roughly 597,000 million kilograms worth $1.08 million in 2009.

Over the years, Agbogbloshie has become a place of urban economic possibility for many of Ghana's most marginalized people, especially migrant Dagomba laborers from Ghana's Northern Region, which during Ghana's colonial period was a common source of slave extraction.[5] Beyond an apocalyptic and decomposing digital death zone, Agbogbloshie is for many a busy center of socioeconomic vitality. The sounds of the scrap market are a constant reminder of this social fact. It is also a loud environment of ongoing material breakdown. The market soundscape is a complex mixture. It involves high-pitch metal-on-metal—or, to be more exact, the labor sounds of "human-metal" (Mbembe 2017, 180). The clanging and banging of workers pulling metal scrap from the beds of trucks is mixed with heated business negotiations, grazing goats baaing, the upbeat rhythms of Highlife—Ghana's popular musical genre—and the dimmer crackling of e-waste incineration. Groups of workers in different sectors of Agbogbloshie can be found burning bundles of copper wire (see Figure I.1), a sure way to make a slim daily wage. This wire fire labor involves the salvaging of copper scraps, a practice that links Ghana's working poor to the dirty and elusive global copper supply chain. These scrap metal scavengers and miners have direct experience with the heat of this toxic supply chain, direct experience with the bodily risks of e-waste burning itself. Their copper scrap extraction labor involves direct bodily exposure to toxic fumes emitted from the burning of wires enwrapped in plastic insulation. Copper extraction involves burning this excess plastic, a practice of "unregulated" waste management experienced in many countries in the Global South engaged in "quick and dirty reclamation of heavy metals" (Slade 2006, 279).[6] Doing this highly toxic incineration work puts countless people (mostly Black bodies) on the toxic front lines of a fomenting circular economy known for its non-transparency and legal invisibility (Corvellec

---

[5] In fact, among Akan (Twi)–speaking peoples in the central coastal region of Ghana—an original hub of the transatlantic slave trade—the term for "bought person" (ɔdɔnkɔ) became synonymous with northerner (Holsey 2008, 2013). But, according to others I have asked about the translation, the term is closest to "reserved laborer."

[6] As Slade (2006, 279) notes, in addition to the popularized Guiyu site in China, "In India, Pakistan, and Bangladesh, unregulated facilities burn excess plastic waste around the clock, pumping PBDE [polybrominated diphenyl ether] and dioxin-laden fumes into the air."

Figure I.1. Copper wire burners in Agbogbloshie. Photo by author.

et al. 2020; Picard and Beigi 2020), two entangled logics of supply chain capitalism more broadly (Tsing 2009).

Aside from a few small plots of land set aside for growing staple crops, the land in Agbogbloshie is mostly plantless. Hardwood plants are almost nowhere in sight, so the primary fuel source for the wire fires includes anything from junked tires and foam insulation from refrigerators to any available and ignitable petrochemical material. At the burn site, workers usually have a pile of this fire fuel nearby, ready to use whenever copper wire bundles arrive. These wire bundles range in size and are worth between 8 and 10 Ghana cedi (roughly between $2.00 and $2.50), though global metal markets directly impact the flexibility of local copper prices.[7] It takes workers about 10–15 minutes to burn off all the plastic insulation. To speed up the burning, workers flip the torched bundles every minute or two to increase oxygen to stoke the fire. If an experienced burner witnesses another burner burning in

[7] As Grossman (2006, 24) notes, another analytic challenge in the study of copper markets is that "figures concerning copper are often not publicly available and are highly proprietary, particularly in the case of super-competitive high-tech manufacturers." See Akese (2014) for a more detailed study of e-waste price realization in Ghana.

an inefficient way, they will commonly intervene and tease them about not doing it correctly. After the wire bundle is completely burned, the copper wire is bagged and sold to scrap metal dealers, usually Nigerians, who then sell the copper for export in Tema, the planned industrial port in the greater Accra area several miles east of the city center.[8] Workers tell me all the copper goes to China and India, but, according to international trade experts, copper scrap in Ghana is also exported to Italy, the United Arab Emirates, and South Korea, which was the top export destination in 2019.

Some groups of workers in Agbogbloshie are searching for aluminum, the most abundant metal, by mass, found in the Earth's crust. Once heated and broken down, sheets of aluminum are packed tightly into large white bags. In the summer of 2016, aluminum was selling for 1.7 Ghana cedi (about 40 cents) per pound. A full bag weighs about 25–30 pounds, and the profits from this toxic labor are distributed among the workers in a highly structured way. In fact, according to one report on informal e-waste labor in Africa, daily revenues typically range between US$0.22 and US$9.50, and most of these workers in Ghana and Nigeria are making less than US$1.25 per day, which is below the official international poverty line. Along with irregularities in daily incomes and health struggles that can impact earnings, a system of wealth distribution is also at play among workers. A variety of factors impact these local socioeconomic processes. Especially critical are how long they have been working in Agbogbloshie, their social ties and connections with local chiefs or chiefs in their villages in northern Ghana that have close kin working in the scrapyard, or whether they are married or have children. Furthermore, workers in Agbogbloshie who oversee the scrap metal scales are notorious for taking an unfair cut by tinkering with their scales, which creates confusion about the value of the bag of copper or aluminum collected (Akese 2014). In this way, the scrap market is not devoid of scrap scams, and disputes between workers are an everyday intersubjective occurrence. The crackling of copper wire fires and the human–metal soundscape dominating this urban discardscape are equally symbolic of Agbogbloshie's broader social and environmental materiality.

All of this, and more, makes up Agbogbloshie's complex cultural milieu. In this book, my ethnographic focus is on the lived experiences of a small group of migrant laborers from Ghana's Northern Region who work in

---

[8] Tema has also been a focal point of urban waste management critiques and anthropologies of waste (see Chalfin 2014, 2017).

Agbogbloshie burning e-scrap to extract copper. "We burn to live," as one worker told me. Turning an ethnographic eye on this copper scrap labor experience, one begins to see the real problems and moral consequences of making toxic burning a subject (or target) of green neoliberal intervention. *Burning Matters* critically engages the practices and logics of "do good" (Lashaw, Vannier, and Sampson 2017) extrastate institutions and transnational organizations that make burning bad and turn Agbogbloshie into a space reduced to toxic negativity and a work environment in need of decontamination, formalized innovation, and "best practices." These technoscience interventions promote global environmental health and encourage urban waste management reform as much as they extend an economized ethos of green and clean "planetary improvement" (Goldstein 2018). But urban waste management and pollution control in Ghana, as others have pointed out, are far from narrow sociomaterial challenges with apolitical consequences (Chalfin 2017). In response to interventions taking a compressed technocratic angle, this book attempts an alternative path aimed at opening up Agbogbloshie to an analysis centered on cultural, environmental, and political complexity, a complexity that is too often lost and overshadowed by runaway myths and inaccuracies (van der Velden and Oteng-Ababio 2019; van der Velden 2019). For example, to even begin to fully understand the urban arrangement of Agbogbloshie, one needs to acknowledge certain social and environmental governance dynamics in this area of Accra.

Most people who work in Agbogbloshie live in a large residential community or "slum"[9] known as Old Fadama, which is located about 100 meters across the Odaw River to the east of the main Agbogbloshie scrap market. Old Fadama, one of nearly 76 reported slums in Accra alone (Ernest 2020), contains little to no electricity or running water; and local governance is elusive, even despite the establishment of a de facto local government organization called the Old Fadama Development Association. As noted by Stacey and Lund (2016, 595), in Old Fadama "different actors claim a variety of rights and services—to land, to building and settlement, to schooling for their children, to health services, to conflict resolution, and to security, to name but the most basic." This asymmetry in rights leads to some residents experiencing

---

[9] As many have pointed out, "slum" is a problematic term that anchors urban poverty to all things negative, even turning complex places into places of exploitative branding. The recent expansion of a slum tourism industry sharply illustrates this exploitative dimension of the term (Dürr and Jaffee 2012). Even organizations like Slum Dwellers International have adapted their concepts, now preferring the phrasing "shack/slum."

more continuous harassment and "formal" eviction than others (Morrison 2017). Moreover, portrayals of Agbogbloshie as an apocalyptic wasteland and "the world's largest e-waste dump" only generate more justification for slum evictions by Ghanaian authorities. Additionally, these evictions reinforce the ambiguity of governance and authority within Old Fadama, even though many blame Accra's political and administrative authority, the Accra Metropolitan Assembly, an urban agency whose mission is "to improve the quality of life of the people of the city of Accra, especially the poor, vulnerable and excluded, by providing and maintaining basic services and facilities in the areas of education, health, sanitation and other social amenities, in the context of discipline, a sense of urgency and commitment to excellence." But, perhaps most importantly, eviction and displacement in Old Fadama are further complicated by the fact that a local land management authority is not clearly defined and "reciprocal dynamics are not established between a singular coherent authority and a population, regarding the provision of services" (Stacey and Lund 2016, 596).[10] But while urban governance in this area of Accra remains elusive and land rights remain an everyday source of dispute, it is well known that settlement in what is now Agbogbloshie and Old Fadama began during Ghana's colonial period. These first settlers were primarily Housa traders, followed by the "railway community," which was home to predominantly rail line workers. Most people talk about Agbogbloshie and Old Fadama as the same general area, but despite place-name correctness and complexities, the scrap market and site of e-waste recycling activity is defined to a particular area (see Map I.2). Currently, the Agbogbloshie and Old Fadama area is home to roughly 100,000 people and is considered a strategic economic hub for Ghana.

Over the years, Agbogbloshie has become a centerpiece of global e-waste pollution intervention,[11] a place of "NGOization and experimentality" (Murphy 2017, 90). In 2014, Agbogbloshie became the site of a "model" e-waste recycling center built to make e-waste recycling work safer and more environmentally friendly. The primary mission is captured in a simple declaration: "Eliminate Burning at Agbogbloshie." With support from a variety of government and nongovernment agencies, the new recycling facility,

---

[10] Under these conditions, "authority is not exercised by government as an extension of formal state law, but emerges though everyday negotiations of claims to services in return for obligations" (Stacey and Lund 2016, 296). See also Afenah 2012.

[11] There is no shortage of research pointing out the ways in which African "crisis" discourse has led to increases in "experimental" interventions, turning postcolonial Africa into an experimental zone to test "crisis" solutions and remedies (Rottenburg 2009; Nguyen 2009; Prince and Marshland 2014).

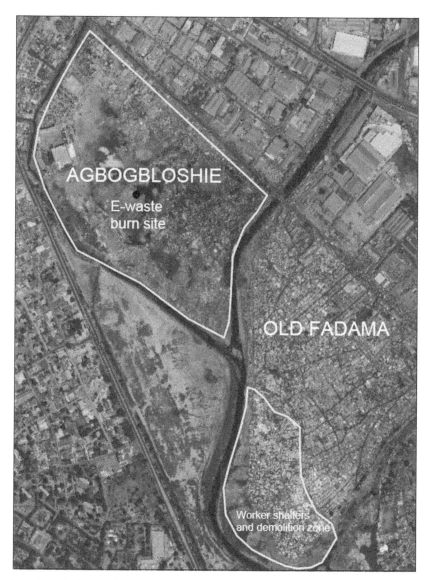

Map I.2. Agbogbloshie and Old Fadama.

recently named the Agbogbloshie Recycling Center (ARC), aims to decontaminate Agbogbloshie but also reinforces the effort to turn this
scrap market into an "innovation hub" (Bardsley 2019). The partnering
agencies and organizations behind the ARC project include the European
Commission, the United Nations Industrial Development Organization,

the Global Alliance for Health and Pollution, the Ministry of Environment, Ghana's Environmental Protection Agency, Ghana Health Services, the National Youth Authority, Pure Earth, Green Advocacy Ghana, and the Greater Accra Scrap Dealers Association. In short, the ARC aims to reduce the health risks of electronic cable burning—only one of many sources of air pollution in Agbogbloshie—by using automated granulators to strip coated cables and wires of various sizes containing copper and other valuable, yet toxic, materials. Some of the e-waste accumulating in the scrap market is large-diameter electrical cables—aluminum-based utility lines—from either Ghana's electrical utility company or neighboring countries, like Burkina Faso to the north. But the bundles of wire that the majority of burners salvage and scorch are much smaller in diameter and are collected from small household electronics, junked cars, buses, and trucks (Akese 2014). These other forms of discard are a combination of electronic and electrical wastes, tech discards sourced from domestic machinery containing copper wires.

There is no question that electronic discard is accumulating on a global scale (Parajuly et al. 2019; Lepawsky 2018; United Nations Environment Programme 2015). A useful factoid that might help capture the magnitude of the problem is that "The amount of e-waste generated each year is equivalent to the weight of 5,000 Eiffel Towers" (Parajuly et al. 2019, 9). Each year roughly 50 million tons of e-waste are generated worldwide, and most, if not all, of this digital discard contains plastics that enwrap electrical wires and other components. These plastics, we now know, are a dominant disposable material that is a widely recognized global environmental challenge.[12] For example, based on historical trends, global cumulative plastic waste generation is expected to reach over 25,000 million metric tons by 2050 (Geyer, Jambeck, and Law 2017). But e-waste, like other wastes, has many signatures and involves careful attention to broader systems of disposal, salvage, and reuse (Isenhour and Reno 2019; Fredericks 2018; Millar 2018). So while it is important to acknowledge e-waste as one of the fastest-growing hazardous waste streams in the world, the picture isn't close to complete without also noting that the e-waste "recycling"[13] industry is an $18 billion trade industry, up from $700 million in 2003 (Kaza et al. 2018; United Nations Environment

---

[12] According to Geyer, Jambeck, and Law (2017), plastic waste generation has been increasing in key exporting countries like Germany, where there was an increase of 3.9% between 2015 and 2017, and the United States, where there was an estimated increase of 12% between 2015 and 2018.

[13] As Slade (2006, 279) notes, " 'Recycling' is a word with unexamined positive connotations" that "obscure[s] a host of ills."

Programme 2015). Agbogbloshie exists within this general "worlding" of e-waste that is uneven (Lepawsky 2018; Parajuly et al. 2019). E-waste is tethered to a global electronics industry that doesn't show any signs of a diminishing supply, especially in a consumer market dominated by corporate-driven "made to break" disposability logics (Slade 2006; Byster and Smith 2006; Suarez-Villa 2009). Moreover, while the global electronics industry shows signs of creatively inserting "extended producer responsibility"[14] policies at all stages of electronics design, manufacturing, assembly, consumption, and disposal, many in the industry know that "planned obsolescence historically has been a key feature of the market-oriented, production-consumption life cycle" (Byster and Tu 2006, 201).[15]

While most electronics are still disposed of by conventional means (i.e., in a landfill or incinerator), an increasing number of these discards are recycled and reused for the purpose of extracting "secondary raw materials," especially valuable metals like copper and aluminum. In the United States in the 1970s, for example, electronic discard was a source of metal extraction, rather than a focus of waste management (Kleespies, Bennetts, and Henrie 1970; Dannenberg, Maurice, and Potter 1973). In fact, one can't actually separate e-waste trades from emerging circular economies and global "circuits of extraction" (Arboleda 2018, 2020). As recent extraction[16] studies note, "circuits of extraction not only entail the physical distribution of raw materials across logistical space (i.e. their organization as supply chains)" but also involve their relationship to "other realms of economic and territorial life" (Arboleda 2020). E-waste and scrap metal reuse economies and ecologies are extraction worlds with direct and indirect global supply-chain linkages and logics.[17] For example, there are significant economic rewards of e-waste recovery, reuse,

---

[14] For a review of extended producer responsibility (EPR) campaigns in the United States, see Raphael and Smith (2006); and for a review of EPR legislation in Japan and Sweden, see Tojo (2006). Also see Lepawsky (2012).

[15] The Silicon Valley Toxics Coalition, a California-based organization, was one of the first international advocacy groups to expose the global dynamics of electronics industry pollution and e-waste (see Keirsten and Michael 1999).

[16] Following Sze (2020, 142), extraction refers to "The unsustainable use of natural resources (oil, mining, etc.), often shaped by race/racism and colonial histories/neocolonial policies."

[17] Extractivism has fast become an expansionary concept and focus of ethnography (Jacka 2018; Dobler and Kesselring 2019; Jalbert et al. 2017; Pearson 2017). As Arboleda (2020) notes, much of this extraction theory "is premised on the idea that some of the dynamics and logics of primary commodity production are becoming rapidly extended to other domains of socio-economic activity such as finance, real estate, logistics, and the platform economy. Extractive processes, according to these authors, provide important analytical insights for elucidating the role of rent, primitive accumulation, and extra-economic force under contemporary capitalism, especially since the Great Recession of 2008."

and re-commodification of increasingly valuable metals (Lepawsky 2018; Reddy 2016), metals including, but not limited to, gold, copper, nickel, as well as rare materials like indium, palladium, and coltan.[18]

Copper, Toxic Supply Chains, and Technocapital Ruination

We live in a time when "planetary entanglement and technological acceleration" (Mbembe 2019) matters. Electronic things, technologies, devices, and equipment are profoundly shaping all domains of late modern life, from science, communication, healthcare, transportation, governance, energy, and entertainment to business and education. Amidst e-banking, e-commerce, e-voting, e-science, e-gaming, e-smoking, e-warfare, e-surveillance, e-policing, and so on, it is not hyperbole to consider the "human" itself in *Homo electronicus* and digital terms (Mitchell 2005; Helmreich 1998).[19] The fact is, our warming world of wires and electrical equipment, from undersea cables layering 750,000 miles of ocean floor (Satariano 2019) to power lines weaving through tree-lined city streets, has made all of this tech acceleration the new normal. Our current electronics era—the digital "Capitalocene" (Moore 2016) one might argue—is a time marked by planetary-scale transformations in tech production, electronics consumption, and "wireless" modernity (Crocker and Chiveralls 2019; Lipovetsky 2011; Schor 1998; Fishbein 2001). For example, nearly half the world's population are internet users, and with a growing global middle class spending more each year on electronics, not only does the majority of the global population now have access to mobile networks and services but many, including my Ghanaian interlocutors, own more than one information and communication technology device (Baldé et al. 2017, 4). Additionally, even amidst the global economic slowdown and structural stagnation from the COVID-19 pandemic, profit-driven Big Tech firms (e.g., Google, Amazon, Facebook, Zoom, and Microsoft) are seizing the opportunity to increase capital and power by expanding their consumer

[18] Coltan, a key material used for electronic capacitors, has been the subject of toxic mining reporting in Africa (Harden 2001).

[19] Additionally, one can't deny that alongside praise for an ever-growing "internet of things" (Greengard 2015) and global interconnection are racist algorithmic technosciences (Benjamin 2019a, 2019b; Browne 2015; Wang 2018).

markets (Zuboff 2019, 2021; Klein 2020a, 2020b; Davis 2020; Blakeley 2020; Panitch and Albo 2020; Precarity Lab 2020).[20]

These broader technocapital trends and pandemic profiteering patterns rely on material supply chains, and copper is a central metal to this industry. Global trades and flows of e-waste are equally streams of capital tightly bound to the planetary mining of metals (Labban 2014). Copper extracted from electronic discard reminds us that copper supply chains involve a kind of plasticity, vibrancy, and vital materiality (Bennet 2010; Gregson and Crang 2010). Copper and copper scrap, in this way, is a material that is both "toxic and life-giving" (Millar 2018, 32).

According to the Copper Development Association (CDA), the market development, engineering, and information services arm of the US copper industry, copper is refueled pandemic "business opportunity." Everywhere under our noses, copper is a vital commodity and infrastructure that supports, connects, and wires our late modern lives.[21] According to the CDA, the call for copper is louder than ever:

> Copper keeps the lights on, the machines working, and the trucks moving. Copper ensures the reliable delivery of distributed energy from onsite solar panels, wind generation, ancillary power sources, and large-scale battery storage to keep critical facilities functioning even when grid power is lost. . . . Copper powers the data centers, cell towers, servers and all the equipment necessary for us to communicate and function in a digital world. Computers, cell phones, televisions, radios all rely on copper products to deliver and manage power and data. . . . Copper and many of its alloys (over 800) are naturally antimicrobial and are used to manufacture self-sanitizing surfaces that can assist in slowing the spread of infectious diseases, including COVID-19, in current and future outbreaks. . . . And, while credit cards are plastic and mobile and non-touch payments can be made from our phones and devices, all of those systems and the data generated is powered and delivered through copper and copper alloy parts (Copper Development Association 2020).

---

[20] Big Tech firms are even banking on the COVID-19 pandemic to justify the social–corporate control of the world's "useless" laborers (the caring class), a situation that will only deepen global inequalities (Shiva 2020). This is also pushing the limits of neoliberalism and digital capitalism (Panitch and Albo 2020; see also Huws 1999).

[21] As noted by Schipper et al. (2018), current demand for copper is largely driven by applications of copper in the infrastructure and building construction industry, which makes up approximately 72% of global copper consumed.

We also can't ignore recent "circular" economic interests emerging in the global e-scrap and recycling industry. In fact, a 2019 report published by the Unites Nations E-Waste Coalition and the Platform for Accelerating the Circular Economy is raising awareness and industry excitement about creating a "new circular vision for electronics": "If developed in the right way, employing a circular economy for the electronics and e waste sector could create millions of jobs worldwide. Some may be in low-paid and low-skilled work as more e-waste is reclaimed into the system, but over time, this will change with a wide range of job opportunities emerging" (United Nations E-Waste Coalition 2019, 18). But despite this emerging circular economy optimism, we also know that problematic "holes" and waste politics exist in emerging circular economy thinking, practice, and design (Basel Action Network 2018; O'Donnell and Pranger 2020). This is especially true in the Global South, where some of the most toxic e-waste extraction and recycling practices occur. In these economically disadvantaged places in the Global South, any e-waste struggle is folded into ongoing colonial and postcolonial configurations and forms of structural inequality that challenge efforts to reduce extreme poverty, expand social inclusion, and meet sustainable development goals (Gutberlet and Carenzo 2020; Gutberlet et al. 2017; Velis 2017). This raises numerous concerns about the uneven geographies of contemporary e-waste recycling and metal reuse economies in postcolonies like Ghana. Current e-waste recycling labor in this sub-Saharan African nation amounts to informal "dirty" work (Zimring 2015; Fredericks 2018; Moreno-Tejada 2020)[22] in a postcolonial Africa involving "complex colonial matrices of power" (Ndlovu-Gatsheni 2014; Mbembe 2001; Mignolo and Walsh 2018). So, it is near impossible to isolate global e-waste management and policy challenges, or any other waste trade for that matter, from the actual dynamics of nation-state formation and capital extraction that "are actively remaking urban, financial, and logistical landscapes" (Arboleda 2020) and concurrently producing toxic environments.

While many metal market economists and business enthusiasts forecast a significant copper boom resulting from accelerations in digitalization and green technologies, as well signaling copper's central role in sustaining energy futures "given the expected increase of copper-intensive low carbon

---

[22] For a US-based history of scrap metal labor and its relationship to environmental racism and justice, especially the role of waste labor justice concerns within civil rights movement and action, see Zimring (2015).

energy and electrification of transport technologies" (Schipper et al. 2018, 29),[23] copper mining and production have a noxious environmental history (Hong et al. 1996; Nriagu 1990, 1994, 1996). For example, the history of toxic emissions from copper mining and smelting operations is staggering. From Butte, Montana, to Jiangxi, China, the history of copper mining and production has always been one of atmospheric contamination. In fact, well before even the Industrial Revolution, copper as well as lead were signature metallic pollutants.[24] But unraveling the planetary history and dynamics of copper trades, copper applications, and copper's environmental history is not what *Burning Matters* seeks to do. Nor is it my intention to dwell too extensively on the exploitative and transnational details of supply-chain capitalism and its global environmental destruction. Instead, as an e-waste ethnography, the book explores the situated and nitty-gritty matters of concern for migrant laborers who burn e-waste to salvage copper. What concerns me most is how e-waste labor and life in Agbogbloshie calls for attention[25] to other "burning truths" (Minter 2016) emerging in this West African e-wasteland.

E-Wastelanding and Global Environmental Justice

Making sense of e-waste in Ghana, as elsewhere, requires careful consideration for how e-waste is conceptualized and how e-waste has evolved in relation to broader waste and waste management challenges and policies informing the proliferation of e-waste on a global scale. As discard studies scholars have pointed out, e-waste—what has been variously termed discarded electronic and electrical devices, waste electrical and electronic equipment, or discarded electrical and electronic equipment—is actually a

---

[23] Some recent reports are even highlighting and projecting how this copper boom will play out as governments around the world respond to the COVID-19 pandemic (see Taylor 2020; Fernandez-Stark, Bamber, and Walter 2020).

[24] Ice core samples from Greenland, for example, show that "emissions of copper into the atmosphere surged twice before the Industrial Revolution, once after the introduction of coinage in the ancient Mediterranean, and once with the intense marketization of the Chinese economy—and burgeoning copper production—of the Song dynasty (960–1279 A.D.)" (McNeil 2000, 56). McNeil adds "Inefficient smelting technology sent as much as 15 percent of smelted copper into the air. Total copper emissions in the Roman and Song eras came to about a tenth of those of the 1990s, even though copper production was less than a hundredth of modern levels. Regional, indeed hemispheric, air pollution is about 2,500 years old, and—for copper emissions at least—was as great in Roman and Song times as at any time before 1750" (McNeil 2000, 56).

[25] "Attention" comes from the Latin word *ad-tendere*, which means to stretch (*tendere*) toward (*ad*). See Ingold (2017) for a deeper dive into *attentive* anthropology.

rather complicated term to define. Any definition calls for some recognition of the term's "inherent arbitrariness" and the fact that e-waste involves "material heterogeneity, complexity, potential toxicity, and the degree to which it might be reusable or repairable" (Lepawsky n.d.). Since waste is "intrinsically, profoundly a matter of materiality" (Gregson and Crang 2010, 1026), it has become a source of excitement in anthropology and other social sciences (Reno 2015). Like pollution, waste can also "be used as a lens through which to dissect the social and cultural intricacies of the urban environment, space, and power" (Jaffe 2016, 15). Additionally, waste management itself is now considered an actual "social relationship, rather than an unmediated connection that individual consumers have with an abstract and impersonal Nature" (Reno 2016, 218). Recently, hazardous waste—which includes e-waste—has sparked discussion of inequities in global waste distribution and the environmental racism and violence underpinning what has been described as systematic settler colonial "wastelanding" (Voyles 2015). For example, many scholars and activists engaged in waste and waste management politics have elaborated the capitalist production–waste–environmental justice connection (Bullard 1990; Mohai and Bryant 1992; Szasz 1994). These critiques see capitalist growth as predicated on the unequal and exploitative distribution of benefits and costs, with privileged populations reaping the benefits (whether in the form of profits or consumption) and marginalized groups receiving a disproportionate share of the toxic burdens. With the acceleration of technocapitalism and corporate globalization in the 1990s and early 2000s, these dynamics have taken an aggressive global turn, inspiring scholars to focus more attention on international environmental justice issues through the direct and indirect trade in hazardous wastes (Clapp 2001; Pellow 2007; Faber 2008). Attention to planetary e-wastelanding extends and recycles some of the same toxic concerns and politics highlighted in the wake of the "environmental decade" of the late 1960s, the anti-toxics movement in the 1970s, and the rise of the environmental justice movement[26] in the 1980s, when costs for disposing of hazardous wastes in the United States rose dramatically.

In the absence of regulatory oversight and a decrease in transportation costs, the toxics industry began searching beyond national borders

---

[26] Following Sze (2020, 141), we can think of the environmental justice movement as "A social movement to further policy and cultural changes that support social justice and environmentalism, broadly defined, connecting issues of race, class, indigeneity, gender, citizenship/nation-state, and sexuality with environmental equity."

for cheaper sites to dispose of these wastes. Lax or nonexistent regulations and desperate financial need in "neoliberal welfare states" (Ferguson 2015) of the Global South have not only led to "shadow" economies and labor conditions[27] but also turned certain Global South spaces into susceptible dumping grounds for externalized hazards (Armiero 2021; Müller 2019; Borowy 2019).[28] African e-wasteland economies, from Accra to Lagos to now Dar es Salaam, Tanzania (Yee 2019), are on the upswing across the continent. A pan-African e-waste "crisis" is in the making, and Agbogbloshie has become a centerpiece narrative augmented by ongoing social media attention and research-driven intervention. Agbogbloshie is equally entrenched in a global copper supply chain and planetary system of raw material extraction that is "constituted through racial and spatial politics that render certain bodies and landscapes pollutable" (Voyles 2015, 10).[29] During the 1980s and into the 1990s, this economic and political "push" against toxic polluters in the United States was combined with the "pull" of globalization processes—namely, the decreasing costs of transportation and communication and the increasing receptivity of countries in the Global South of such wastes, with Africa playing a major role. The growth in importation of these wastes was partly the outcome of devolving state support, deregulation, and a lack of cash flow within countries struggling with the multiple burdens of failed structural adjustment policies (Aryeetey and Tarp 2000; Pellow 2007; Clapp 2001; Ferguson 2006).[30] Africa became a focused target for such "stabilization" policies in the 1980s and 1990s but instead has become a glaring example of destructive and stagnating post–World War II economic "development" policy. These trade policies largely favored export-based wealth extraction and tremendous debt inflation, as is the case in Ghana, where three primary commodities—gold, cocoa, and oil—keep this highly indebted and depleted economy in motion. These commodities, when combined, make up over 80% of Ghana's exports, but revenue generated from these exports isn't enough to pay off loans, leading to repeated reliance on loans from private,

---

[27] See Rosen (2020) for a detailed ethnographic account of "shadow rule" among Ghana's small-scale gold miners.

[28] As Borowy (2019, 12) notes, "Waste reflects not only the developmental stage of a given society, but also developmental asymmetries between different regions. All but two of the 50 largest dumpsites worldwide are located in low-income countries in the Global South."

[29] If searching for a more fitting term, one might call this the "racial Capitalocene" epoch (Vergés 2017).

[30] As Ferguson reminds us, "Instead of economic recovery, the structural-adjustment era has seen the lowest rates of economic growth ever recorded in Africa (actually negative, in many cases), along with increasing inequality and marginalization" (2006, 11).

multilateral, and bilateral lending agencies (Jones 2016).[31] So, along with development-based "stabilization," commodity-dependent economies, and enduring hyper-debt came new structures and conditions of "crisis," a heuristic which has continually functioned to order and control framings and responses to ruination and destabilization, especially in vital times of planetary urgency and climate disturbance (Masco 2016, 573; Chakrabarty 2019).

But e-waste crisis talk in Ghana also connects to the histories (and ironies) of global waste trade policies which reinforce the so-called motion in the system (Trouillot 1982), motion involving technocapital flows and e-waste planetarity. The actual success of anti-toxics and environmental justice movements in the 1970s and 1980s in the Global North—while certainly not their intended goal—resulted in citizen responses like "not in my back yard," which ultimately increased the global circulation and exportation of unwanted hazards (e.g., countless chemical byproducts of production and postconsumer wastes like e-waste) to poor communities across the Global South that have been systematically excluded from the sociomaterial benefits of technocapital development. Hope for a change to this chaotic system of hazards exportation came with the development of the Basel Convention on the Transboundary Movement of Hazardous Waste, which had its first meeting in 1989 and entered into force in 1992. As described in more detail in Chapter 1, it is important to note that this was the first global convention designed to regulate the movements of hazardous waste—including e-wastes—across national borders. Moreover, it helped spawn regional versions such as the Bamako Convention in Africa. Enacted in 1998, the Bamako Convention is a treaty of African nations prohibiting the import into Africa of any hazardous waste, including radioactive waste. The impetus for the Bamako Convention arose also from "the failure of the Basel Convention to prohibit trade of hazardous waste to less developed countries" and "The realization that many developed nations were exporting toxic wastes to Africa." The Bamako Convention, in this sense, was a response to "toxic colonialism," a term deployed in early e-waste studies (Brigden et al. 2008; Basel Action Network 2002, 2005; Pellow 2007) that "portrays developing countries as remaining in subordinate positions with regard to e-waste

---

[31] One comprehensive Ghana debt study notes that by the end of 2015, Ghana had $17.4 billion in private loan debt, $5.5 billion in multilateral loan debt, and $4.9 billion in bilateral loan debt (Jones 2016).

dumping, and as a result experiencing environmental injustice as a new kind of colonialism" (Akese 2014, 36).[32]

In 2012, a report was released by the Basel Convention's E-Waste Africa Program, a program with a primary aim to enhance "the environmental governance of e-wastes and creating favorable social and economic conditions for partnerships and small businesses in the recycling sector in Africa" (Basel Convention n.d.). This report highlights how e-waste recycling and refurbishing activities in West Africa have boomed in recent years as a result of lax importation policies (Schluep et al. 2012).[33] But what is often overlooked is how even though the global trade of "good-quality" and functional electronic discard expands the benefits and material "goods" of technocapitalism, it also sustains and exacerbates the exhaustive and ruinous consequences of toxic colonialism. Of course, the toxic injustices of e-waste trades and their manifestations in West Africa do not exist in a historic vacuum or a history devoid of violent patterns of global capitalism.

For many of those working in Agbogbloshie, copper capitalism is a force of labor. Workers extract, recover, and sell copper, a metal of global demand that is ubiquitous in almost any development project requiring electricity (Gupta 2015; Boyer 2015). In this way, burning e-waste in Ghana to extract copper is a topic that cuts deep into broader politics of toxic supply-chain capitalism, a system of economic domination and exploitation shaped by ongoing legacies of violent colonialism, White supremacy, and slavery (Horne 2020). This metal, in the form of copper bars, operated as a prominent currency in Africa for centuries, especially during the colonial period when slaves were used to repay debts. As anthropologist David Graeber explains, copper as currency was integral: "The Atlantic Slave Trade as a whole was a gigantic network of credit arrangements. . . . On arrival, European traders would negotiate the value of their cargoes [carrying slaves, sugar, tobacco, iron, linens, etc.] in the copper bars that served as the currency of the port" (2011, 150–151). In light of this copper–slave capitalism relationship, it is now well known that African bodies and African slave labor were paramount to the expansion of plantation and extraction economies in the Atlantic

---

[32] Liboiron (2018a, 2018b) offers a detailed discussion of the discursive origins of "waste colonialism," but Povinelli (2019, 2) sums it up in the following way: "Lifted up, lifted out, anthropos was claimed to be different from and superior to all other forms of existence. But this anthropos is not Man. It is a toxic imaginary brewed out of specific colonial and capitalist sociality."

[33] As Burrell (2012) notes, much of Ghana's early e-waste management challenges are directly linked to the growth of internet cafes in the late 1990s, which led to increased importations of second-hand computers from the Global North.

region (Klooster and Padula 2005).[34] So we see an instant entanglement of toxic copper and African extraction and slaving in the formation of a global-ized "racial capitalism" (Robinson 1983).

Anchoring *Burning Matters* is a concerted effort to explicitly and cau-tiously humanize and make audible what is already a focus of many impor-tant global waste studies. It adds flesh to studies concentrating on mapping global material flows, trade policies and negotiations, and even recent cir-cular economic analysis attending to "the embodied carework of tinkering, repairing and tending to materials" (Isenhour and Reno 2019, 1) and no-table inequalities in e-waste recycling work in particular (Schulz and Lora-Wainwright 2019). As one of my Ghanaian friends put it, "Here in Ghana, they may not tackle the issue from the bottom. They will just go straight to the top to tackle it." What is going on at the "bottom," at the ground level, in Agbogbloshie are the bare consequences of hazardous waste trades, flows, and materials embedded in the toxic urban margins. At the "bottom" is also where one finds the accumulation and circulation of "green" interventions that often reinforce power relations that unfortunately do little to serve the needs and interests of Agbogbloshie's most vulnerable and marginalized urban miners (Jaramillo 2020). So we have a situation whereby "in between the Silicon Valleys of the world and struggling communities, one finds all kinds of instrumentations and technological developments, including those informed by an ecological awareness of planetary limits and global climate change" (Escobar 2018, 18). Development-based solutions to Ghana's e-waste problem—as is the case with nearly all poverty alleviation and mit-igation plans practiced in what the World Bank calls a "highly indebted poor country"—often silence the e-waste livelihoods on the bottom of these technocapital developments.[35] Instead, these e-waste livelihoods and truths demand our attention, especially at a time of accelerated electronics consumption.

---

[34] As Mbembe reminds us, "Between the fourteenth and the nineteenth centuries, the spatial ho-rizon of Europe expanded considerably. The Atlantic gradually became the epicenter of a new con-catenation of worlds, the locus of a new planetary consciousness. The shift into the Atlantic followed European attempts at expansion in the Canaries, Madeira, the Azores, and the islands of Cape Verde and culminated in the establishment of a plantation economy dependent on African slave labor" (2017, 40).

[35] As Pierre (2013, 150–151) notes, "In accepting the HIPC [highly indebted poor country] des-ignation, state officials confirmed the depth of Ghana's poverty and its marginalized position within global political and economic hierarchies."

Ethnographic Extensions, Methods, and Positionality

I visited Agbogbloshie for the first time in August 2015, about a year after I completed a book on electronics production, pollution, and risk mitigation in a deindustrialized US town that was the official birthplace of tech giant International Business Machines Corporation (IBM) (Little 2014). Even though it quickly became apparent that e-waste recycling in Agbogbloshie was connected to broader copper supply-chain logics and global e-waste management pursuits, my decision to go to Ghana to conduct ethnographic research was not originally informed by some desire to deepen understandings of global supply chains or electronics industry waste management and consumption politics. Instead, the research sought to explore the situated lived experiences with toxic labor and the micropolitics of pollution control thriving in the shadows of this now planetary state of toxic electronic plunder. Again, the book is an experiment in e-waste anthropology attuned to e-waste embodiment, to an understanding of urban waste that is both imagined (Linder and Meisner 2016) and experienced by situated, laboring, and contaminated bodies that are subject to multifarious toxic struggle and ongoing urban environmental health science and intervention. It is also a situated anthropology grappling with a global challenge, namely the global proliferation and uneven distribution of e-waste and the toxic risks of e-waste recycling labor.

If I had to pinpoint the reasons for turning an ethnographic eye on Agbogbloshie, two reasons stand out. First, given my anthropological interests in high-tech industrial pollution and community contamination, I was inspired by the idea of extending my ethnographic interests to another place, landscape, and sphere of toxic electronics and global environmental health politics.[36] After visiting for the first time, I came to realize that Agbogbloshie was much more than an "electronics graveyard" (Yeebo 2014) or ruin in the life cycle of modern electronics. I quickly learned that e-waste put Agbogbloshie on the global toxics map starting around 2005 and that this place was also integrally linked to a global e-waste trade and scrap metal economy thirsty for recycled copper. I also turned my ethnographic attention to Agbogbloshie because I was interested in exploring and following

---

[36] High-tech pioneer, billionaire, and philanthropist Bill Gates has been a primary funder of global health initiatives; but to date, the Bill and Melinda Gates Foundation has not developed an e-waste environmental health research and advocacy program.

up on e-waste and toxicity[37] matters in a setting that had already attracted other curious social scientists (Fuhriman 2008; Pérez 2014; Akese 2014). But, in all honesty, I was mostly curious to learn about the ways in which toxic e-waste burning at Agbogbloshie lured global environmental health experts, a situation that has led to fluctuating (and often failing) international projects and efforts to control pollution, decontaminate e-waste processing, and make e-waste labor safer.

While doing fieldwork in Agbogbloshie, it became clear that it was not just a popular site in the Global South with rampant end-of-life management and e-waste politics but also a site of precarious pollution control and mitigation logics. To my surprise, Agbogbloshie reminded me of my previous ethnographic work on the toxics mitigation in a US community struggling with high-tech industrial pollution (Little 2014). What I have described elsewhere in "technocapital sacrifice zone" (Little 2016) terms, Agbogbloshie resembles yet another place where waste from the electronics industry was a focal point of intervention; but this time the waste and pollution intervention politics and debates were unfolding in a completely different cultural, political, economic, and environmental context of postcoloniality and neoliberal intervention. In this African e-waste context, I was instead grappling with pollution and risk mitigation politics and practices in a complex urban postcolonial environment where air pollution and especially toxic fire control came to dominate the focus of international and domestic e-waste interventions and global environmental health science interests. In Agbogbloshie, I was confronting not an environment contaminated by the industrial spills of a Fortune 500 corporation, as I was in upstate New York, but instead a toxic environment of extreme poverty and toxic urban marginality that is now enwrapped in copper supply-chain capitalism. Unlike IBM's deindustrialized birthplace, Agbogbloshie resembled an active urban industrial zone fueled by a vibrant metal recycling and reuse economy. Despite these stark differences, these two sacrifice zones of technocapital pollution—one in the Global North and the other in the Global South—involve similar forms of toxics friction and are marked by similar traces of high-tech toxics. So, not unlike the bodily and environmental contamination experienced in

---

[37] "Toxicity" can be defined as "The spread of environmental harm and vulnerability in the depletion economy. Digital industries create toxic matter that spreads and latches onto all kinds of bodies, some more than others. It is both a cultural condition and a material state of being" (Precarity Lab 2020, 100).

US landfills and Superfund sites, life experience in places like Agbogbloshie is often more misunderstood and mysterious than is actually known (Reno 2016). This is perhaps why toxics ethnography matters most, wherever the location and site of struggle might be.

During my repeated visits to Ghana between 2015 and 2018, and to Agbogbloshie in particular, I was always well aware of my Whiteness, of my White American privilege. How workers perceived me during these research trips occupied my inner dialogue each day, and many of my own journal entries from fieldwork illustrate my reflection on and even discomfort with my own positionality and body in this postcolonial research space.[38] But with each fieldwork trip to Agbogbloshie, I grew comfortable with confronting my own Whiteness, a topic that sees little attention in social science research on Ghana's e-waste politics. There are likely many reasons for this, but as Pierre (2013, 177) notes, "Whiteness, though assumed as the power category both in Ghana and globally, remains problematically unpacked and uninterrogated." Taking a more reflexive ethnographic approach, my fieldwork experience in Agbogbloshie has taught me that I can't really make sense of my own social positioning in Agbogbloshie, my positioning as a privileged White male and professor, without taking a hard look at deeper racial politics. I strongly believe Black lives matter and actively joined the 2020 Black Lives Matter protests that ignited across the United States and beyond to reckon with and counter deep structural violence based on racism, police violence, and White supremacy. I am learning to look inward and unpack and interrogate this ethnographic subjectivity, an ongoing reflexive process that gets to the heart of an e-waste studies in line with environmental justice, techno-racial capitalism, and decolonial critique.

Since the 1980s, a growing number of studies have investigated the environmental and occupational health risks for workers involved in both the manufacturing and disposal of electronics (Smith, Sonnenfeld, and Pellow 2006). While social science research directly linking e-waste issues and environmental justice politics exists (Fuhriman 2008; Little and Lucier 2017), there has been little ground-level ethnographic research linking e-waste labor experience and global efforts to decontaminate e-waste recycling

---

[38] Despite fieldwork experiences with discomfort, I did not have the experience of "vulnerable observer" (Behar 1997), which is surely due to my own position as a sizable White American man.

and "green" e-waste management in the Global South. Equally absent are studies exploring the extent to which e-waste interventions actually influence e-waste recyclers' labor, health, and general sense of socioeconomic *precarity.*[39]

Discard studies scholars has emphasized the relations of macro-level processes (e.g., global e-waste trades) with more micro-level concerns and lived experience (e.g., worker experiences) (Lepawsky and Billah 2011; Lepawsky and McNabb 2010; Schulz 2015). Other research has examined the ways in which ineffective implementation of existing global and national environmental policies have adversely impacted the health and safety of e-waste recyclers and local environments (Oteng-Ababio 2010; Yu et al. 2017; Nnorom and Osibanjo 2008). However, research exploring e-waste recycling workers' perspectives and experience in the context of more recent innovations in e-waste policy and e-waste recycling practice is a more limited, but gradually expanding, sphere of scholarship (Kirby and Lora-Wainwright 2015; Schluep et al. 2012; Baldé et al. 2017). This recent scholarship notes that innovations are typically aimed at political, economic, and environmental efforts to convert "informal" e-waste sectors in countries such as Ghana, Nigeria, India, and China into new "formalized" facilities under the guidance of development agencies, NGOs, corporations, and state-supported industries (Lora-Wainwright 2017; Kirby and Lora-Wainwright 2015; Baldé et al. 2017; Schulz 2015). Furthermore, while interest in e-waste community advocacy and environmental justice emerged in Ghana as early as 2008 (Fuhriman 2008), few studies have attempted to directly link these community experiences and perspectives to global environmental justice advocacy discourse and action (Akese and Little 2018; Fuhriman 2008). Using an ethnographically grounded political ecology approach, this book extends these efforts and takes a first step toward rethinking e-wastelanding and e-waste politics in Ghana through a critical and open pyropolitical lens.

---

[39] For recent critical engagements with "precarity" as a conceptual vector to understand lived experiences of contemporary capitalism (and technocapitalism), see Precarity Lab (2020) and Azmanova (2020). Following the definition provided in Precarity Lab (2020, 100), I take "precarity" to mean "a state of being and lived experience of insecurity, loss of control, and unpredictability of one's world"; and I find it useful as a term that "highlights the differential circumstances in which precarity arises, extracting resources from and implicating racial, ethnic, and sexual minorities, women, indigenous people, and migrants who occupy extractive and depleted zones."

A Pyropolitical Ecology of E-Waste

A central argument of this e-waste ethnography is that making sense of toxic e-waste burning in Agbogbloshie calls for a perspective and grammar that avoid reducing Ghana's e-waste problem to urban waste management and discard crisis itself. The toxic pyrocumulous fires emerging from Agbogbloshie contain massive amounts of black carbon and greenhouse gases and are among thousands of harmful fires burning across the globe.[40] Immersed in this truly global pyrocumulous activity, "We all have a stake in the materiality of fire" (Delaporte and Grant 2019; see also Ferguson 2017). In response to this calling and taking a "political-ecological diagnosis" (Little 2014) approach, I argue that a certain kind of political ecology of e-waste is called for, one that involves a more critical analysis of e-waste labor and life that intentionally goes beyond the noxious and thwarting practice of e-waste burning, which has steadily become a primary semiotic anchor point of e-waste politics in Ghana and across the Global South.

Focused on theoretical synergy between political ecology and discard, waste, extraction, and recycling studies,[41] the book couches toxic e-waste politics and complexities in Ghana within a pyropolitical ecology perspective, an approach to e-waste studies that highlights complexity through the critical lens of what I shall call e-waste pyropolitics (or e-pyropolitics). I propose that Agbogbloshie can be thought of as a place of toxic supply-chain "pyropolitics" (Marder 2015), a term that has been literally and metaphorically deployed to rethink political theory and practice in light of a world consumed by "fires, flames, sparks, immolations, incinerations, and burning" (Marder 2015, xiv). For Marder, "the contours of an unjust world" today involve "*how* world destruction, which goes under the name of a globalising world-creation or world-integration, is accomplished. Instead of evaporating into thin air, things are consumed by fire" (2015, xii).[42] It is well known that fire has long been at the center of human culture and history (Pyne 1982, 1994, 1995, 2001, 2012), but thinking of fire in relation to social and

---

[40] For an expanding archive of Earth's fires, see NASA's Fire Information for Resource Management System. Fires containing black carbon are especially concerning as "black carbon not only has impacts on human health, it also affects visibility, harms ecosystems, reduces agricultural productivity and exacerbates global warming . . . black carbon is the most solar energy-absorbing component of particulate matter and can absorb one million times more energy than CO_2," (Cho 2016).

[41] See Nagle (2013), Reno (2014), Isenhour and Reno (2019), Arboleda (2020), and Alexander and Reno (2012).

[42] More recently, fire has even been deployed to make sense of the insanity of Trump's neofascist assault on American democracy (Reich 2020).

environmental conflict and politics shifts critical debate and analysis toward the pyropolitical. While a circulating concept in recent political theory,[43] my interest in this pyropolitical perspective is inspired by recent work in geography, political ecology, and anthropology.

Following political geographers Minor and Boyce (2018), my attention to the pyropolitical stems from a particular interest in recentering fire and fiery extraction in political theory and recent discard studies (Neale and Macdonald 2019; Neale, Zahara, and Smith 2019). E-waste combustion and incineration provide a window into not only "the complex entanglement of state power with the material multiplicities in and through which it is exercised" but also a critical perspective on the interventions activated to respond to these material conditions (Minor and Boyce 2018, 92). Much like Minor and Boyce's call to political geographers to attend more closely to the conceptual utility of the concept of pyropolitics, my call to researchers exploring the interface of electronic discard, toxics mitigation, and environmental justice in a planetary moment of danger (Sze 2020) is to "pay greater attention to combustion and its governance as a constitutive, if heterogeneous, dimension of modernity and collective life" (Minor and Boyce 2018, 92). Attempts to govern e-waste labor and mitigate the burning of e-waste in Agbogbloshie have set in motion an assemblage of relationships with and interventions on toxic fire that ultimately exposes the contemporary materiality and landscape of fire. This situation signals a broader global pattern, a new direction in the current "planetary inferno" (Moore 2019; see also Klein 2019; Thunberg 2019) where the significance of "complex fires" (Pierre-Louis 2019) calls greater attention to emerging political ecologies of incineration and fire.[44] Burning matters are everywhere it seems, from the toxic burn pits of Operation Iraqi Freedom (Poisson et al. 2020)[45] to the burning of toxic

---

[43] As noted in Marder (2015), the term has even been used to critique of alt-right movements inspired by the White nationalist and radical conservative Carl Schmitt. Fire has also been the focus of recent historical studies of slave revolt (Zoellner 2020).

[44] See Hogue (2020) on recent per- and polyfluoroalkyl substance incineration politics and Collins (2008) for a good example of a political ecology of fire approach exploring relations of risk, marginality, and hazard vulnerability.

[45] Used to incinerate garbage and military waste, Poisson et al. (2020, 2) report that

> American forces operated more than 250 burn pits on Joint-Base Balad, Iraq to accommodate the accumulating waste of warfighters and support personnel stationed in the region. An estimated 140 tons of trash was burned daily, with the typical soldier contributing 10 pounds per day . . . a report documented the burning of high quantities of various toxic materials (e.g., plastics, tires, paints, batteries, Styrofoam, medical waste, electronic equipment, pesticides, and human trash) using the jet fuel propellant (JP-8) as an accelerant.

firefighting foam in New York (Harris 2020), from fiery Black Lives Matter protests in cities across the United States in 2020 in response to rampant police brutality[46] to Earth warming climate catastrophe everywhere. We no doubt live in a time when fires, heat, and burning have become more than simple metaphors of struggle. These global realities and ruptures have reanimated and reignited the complex social, political, and environmental materiality of fire and techno burning (Mullaney et al. 2021). In this way, another goal of my pyropolitical ecology perspective is to push the conceptual and geospatial boundaries or limits of toxic e-waste burning.

Doing ethnographic fieldwork on e-waste burning exposes the limits, edges, and geospatial relations of e-waste labor. Ghana's e-waste burners migrate long distances to work in and navigate a toxic market environment in an urban center with non-city, hinterland, linkages. What seems to give e-waste any vibrancy at all is the fact that e-waste extraction and fire actually link peoples and regions within Ghana itself. These urban–hinterland relations—relations informed by kin and social networks as well as by supply-chain capitalism—expose the geospatial limits of urban e-waste burning optics. The circulating toxic-pyro narrative in Agbogbloshie has actual toxic-pyro limits that come to life through an e-waste anthropology attuned to biographies of labor, kin and social networks, and environmental suffering (Auyero and Swistun 2009). In this way, thinking of Agbogbloshie within a pyropolitical ecology milieu encourages a rearticulation of the very "urban" political ecology of e-waste in Ghana as it involves actual relations of toxic supply-chain capitalism reliant on flows of hinterland labor.[47] To this end, my pyropolitical ecology perspective involves an intervention on or rethinking of how e-waste exposes spatialized inequalities that are inscribed on an enflamed and conflicted landscape. Following Broto and Calvet (2020, 282), "A landscape perspective enables an analysis of the production of space with the material and symbolic networks that support capitalism. . . . Conflict is seen as a constitutive feature of landscapes." Unearthing this conflict is a process that ultimately needs to reckon with shared or linked geographies of violence, racism, and colonization that interconnect burning lands, bodies, and justice struggles (Hoover 2019; Ferguson 2017; Perry 2014). In other

---

[46] Fire has long been a centerpiece of urban racial politics and movements in the United States (see Davis and Weiner 2020).

[47] As Brenner and Katsikis (2020, 26) note, "Despite its role in offering powerful scholarly counterpoints to the ideology of the self-propelled city, the bulk of contemporary urban ecological scholarship has confronted the hinterland question only indirectly."

words, places like Flint, Standing Rock, and Agbogbloshie are, as contested cultural–environmental landscapes, actually linked places of struggle. Conflict, it might be said, is always a connector. Conflicts interconnect because they involve shared landscapes of struggle, shared geo-spatial and geo-cultural conflicts that are indeed place-based but also intentionally expansionary in their vision of justice and solidarity struggle. This is precisely why *Burning Matters* attends to frontline Ghanaian experiences of high-tech "wastelanding" (Voyles 2015), which is a conflict-ridden process where power and toxicity, fire and metal, pollution and postcolonial environmental (in)justice meet.

In a move to develop a *pyropolitical ecology of e-wastelanding* in postcolonial Ghana, then, I am rightfully engaged in the reanimation and critique of contemporary forms of toxic colonialism and environmental racism. As Voyles puts it, "Just as race is a discursive technology with often deadly material effects, so too is wastelanding the process by which pollutability is materialized" (2015, 15). Without a doubt, postcolonial Ghana is characterized by ongoing processes of racialization, processes that continue to remind us that Ghanaian society is always set in relationship to and in tension with global White supremacy. Following Pierre (2013), it is important to note that Ghana has since the colonial period been engaged in a trend-setting project of state "racecraft." As she notes, racial identity in Ghana has been conditioned by several ongoing process or definitional projects that call for careful attention. In addition to defining Black racial identity through the transatlantic slave trade, as well as colonial framings of Whiteness in opposition to "native" "tribes," Pierre (2013) identifies two racial projects in Ghana that shape contemporary understandings of racial identity. The first of these is the pan-African fight for independence launched by the first head of state, Kwame Nkrumah, and his cohort, who saw the building of a viable, sovereign postcolonial nation as necessary for combatting a global White supremacy movement that weaponized and racialized fire.[48] Part of this meant infusing the work of national development with Black racial pride and taking a critical stance on neocolonial "aid," which Nkrumah (1965) rightfully insisted was "merely a revolving credit, paid by the neocolonial master, passing through the neocolonial State and returning to the neo-colonial master in the form

---

[48] Early writings on this race-based pyropolitical violence can be found in W. E. B. Du Bois's trilogy *The Black Flame* (Du Bois [1957] 2014, [1959] 2014, [1961] 2014) and for recent reflection on the symbolic-political centering of fire in the radical Black movement, see West and Buschendorf (2014).

of increased profits." Since the 1980s, the Ghanaian state has strategically deployed pan-African rhetoric in aid of market-led development policies (Pierre 2013), even as neoliberal reforms reinforce Ghanaian dependence on and continued exploitation of labor from Ghana's Northern Region, a region now shaped by global e-waste "circuits of extraction" (Arboleda 2020) that break down the boundaries of "racecraft" and ecocraft. All of this is to say that e-waste recycling labor, contamination, and intervention in Ghana take shape in a global territory "tethered to racializing matter" (Yusoff 2018, 14) and global Black ecologies (Hare 1970; Roane and Hosbey 2019; Hosbey, Lloréns, and Roane, forthcoming). Attending to these critical ecologies reanimates how the e-waste toxics crisis and precarious intervention are understood, showing how each "violently reorganizes territories as well as continually perpetuates dramatic social and economic inequalities" (Gómez-Barris 2017, xviii).

My pyropolitical ecology of e-wastelanding would be incomplete without careful attention to the various impacts of a warming and burning planet transformed by ongoing and escalating climatic disturbances. As Ghana faces climate change challenges, the toxic pyrocumulous clouds (or fire clouds) from e-waste burning have quickly become a target of climate change blame. Rapidly contributing to Ghana's overall carbon footprint, pyrocumulous disturbance, it turns out, is a reality not unique to Ghana or anything new. Burning land for farming and to increase agricultural productivity is a cultural practice that for centuries has tainted Africa's air. According to the National Aeronautics and Space Administration, about half of the Earth's annual fires derive from Africa. Extreme changes to precipitation patterns and soil nutrition decline in coastal regions are among the many negative impacts of this pyrocumulous activity (Rasmussen 2015). But aside from these climatic disturbances and their known urban air quality and health impacts (World Health Organization 2016; Cho 2016), Ghana has responded more broadly to global warming concerns through active participation in international climate change negotiations, as well as domestic planning initiatives.[49] For example, Ghana has developed a "sustainable" growth plan that incorporates climate change concerns in various ways, as documented in its Shared Growth and Development Agenda II. However,

---

[49] Amidst domestic waves of climate denial and climate skepticism in the Global North that severely interrupt much needed policy intervention (see Hodgetts and McGravey 2020), global consensus among scientists points to the fact that climate change is real and that significant Earth system changes are absolutely anthropogenic (Powell 2019; Steffen et al. 2018).

Ghana faces serious impediments to "sustainable" growth and the ability to adapt to ongoing climatic changes that have a direct impact on communities across the country. This is, for example, a strong focus of recent climate science in regions most dependent on agrarian and natural resource sector employment, as is the case in most of Ghana and especially in the Northern Region (Tahiru 2019).[50] What has become a systemic trend in the Global South more broadly, "sustainability" in Ghana is "being turned inside out to refer to the sustainability of economic growth and factors sustaining the growth model and to limit environmental regulations that might constrain it" (Black 1999, 140).[51]

As highlighted in this book, these broader economic and environmental realities help expose how, at the grassroots level, e-waste interventions are meshed with "green" neoliberal economic development interests that benefit few in the urban margins and do little to achieve the intended goal of e-waste pollution control and toxics mitigation in Ghana. Additionally, while e-waste accumulation is becoming an operative theme of urban environmental crisis and intervention, climate change impacts on agricultural livelihoods in Ghana's Northern Region are directly influencing domestic e-waste labor migrations. In this sense, it is not just the singular case of toxic fires emitting from Agbogbloshie but a warming planet—the "planetary inferno" (Moore 2019)—that highlights the need to drift toward a broader pyropolitical ecology perspective.

Burning Matters engages a complex tangle of questions. What does e-waste burning in Agbogbloshie teach us about labor and livelihoods? How are these toxic incineration politics linked to circular e-waste economies? How is e-waste burning connected to slum settlement governance and urban marginality and erasure? What is emerging e-waste epidemiology in Agbogbloshie up to, and who benefits? How do e-waste interventions register among poor e-waste laborers, and in what ways do these interventions showcase green neoliberalization in Ghana? Finally, in what ways can pyropolitical ecology and e-waste ethnography advance decolonial discard studies and critical environmental justice approaches in the Global South?

---

[50] Financial limitations matter here as Ghana projects it will need roughly $22.6 billion to carry out the pledge (nationally determined contribution) it made as part of its obligations under the Paris Agreement. Of that money, Ghana projects that $16.3 billion will be required from international donors, which gives some indication of the financial barriers Ghana faces in achieving its "sustainability" goals.

[51] For a recent anthropological critique of the mythical lure of "sustainability," see Checker (2020).

Ultimately, my attention to the grammar of e-pyropolitics is inspired by recent calls for "carving out footholds in stressed and flammable ecological conditions" (Petryna 2018, 573) and to make room for an alternative e-waste optic that connects to critical and global environmental justice theory. In light of this, *Burning Matters* attempts to speak to and hopefully augment e-waste environmental justice politics, especially in a time when global e-waste trades, metal supply chains, urban marginality, and toxic struggles across Africa are on the rise. Extending from the broader political ecology and anthropology of electronics production, discard, and pollution (Little 2014; Schulz 2015), it is my goal, then, to highlight emerging dynamics and complexities of e-waste labor and environmental health justice politics in an African e-wasteland where neoliberal waste management and public health interventions are directly and indirectly inspired by technocapital and circular economic rationalization (Cullen 2017).

At the center of the pyropolitical ecology of e-waste explored here is a more complex narrative addressing micropolitics of pollution control and e-waste economization that ultimately exposes the *political* nature of e-waste intervention. In an attempt to move beyond the dominant toxic prism informing Ghana's e-waste narrative, this book makes the case for a new politics of e-waste burning. It seeks to develop a new kind of footing, perhaps a small ledge to stand on, to explore disciplinary and theoretical synergies. Joining scholarship on waste, recycling, ruination, and rubble (Reno 2016; Alexander and Reno 2012; Gordillo 2014; Voyles 2015), the book seeks to reimagine the political ecology of e-waste and e-waste conceptualization in particular (Gabrys 2011; Kirby and Lora-Wainwright 2015; Pickren 2014; Millington and Lawhon 2019). It explores political-ecological dynamics of e-waste and seeks to develop an e-waste anthropology attuned to critiques of "recycling" (MacBride 2011), pollution–colonialism relations (Liboiron 2021; Voyles 2015), and waste labor more generally (Fredericks 2018; Nagle 2013). Moreover, I approach my critiques of e-waste in Ghana through an intersectional lens that considers the complex and overlapping forces of postcoloniality, environmental injustice, and global "circuits of extraction" (Arboleda 2018, 2020). This intersectional approach also aims to honor the manifold ways in which Agbogbloshie is a lived cultural place of complex connection, as well as a magnet of e-waste "ruination science" (Ureta 2021) and toxic "crisis" rethinking (Masco 2016).

The Chapters in Brief

The book is organized as follows. Chapter 1 provides contextual background on global e-waste policy and politics and emerging "green" neoliberal interventions in Agbogbloshie. It explores NGO interest in Agbogbloshie in general and the practices of one international NGO in particular, Pure Earth, a solutions-based NGO that targets "the most polluted places on Earth." Like other NGOs in Africa and the world over, the "global" is an active ingredient of the intervention ethos of organizations like Pure Earth. The chapter describes the politics and promises of state agencies and NGOs working on urban e-waste projects in Agbogbloshie, especially those aimed at mitigating air pollution and finding solutions to the environmental health crisis there. The chapter explores how this project reflects "green" urban development goals emerging in Ghana and how neoliberal efforts and infrastructures are endorsed and activated to modernize Ghana's scrap metal market economy.

Chapter 2 explores e-waste burning work through the lens of labor migration, city–hinterland connections, and chieftaincy relations and politics. In particular, I focus on the story of one worker's lived experience as a migrant e-waste laborer, husband, father, drummer, and member of a dominant regional chiefdom in northern Ghana. The chapter highlights how e-waste workers navigate urban labor and marginalization in Accra, while at the same time sustaining social ties in northern Ghana where Dagomba chiefdoms hold local and regional political power. The chapter shows how narratives of migration and rural–urban livelihood can expose the integral role of social mobility and movement in e-waste ethnographies more generally.

Chapter 3 explores lived experiences of urban displacement. In particular, I engage and interlink emerging erasure, demolition, and obsolescence logics in Agbogbloshie. I highlight how those working in Agbogbloshie and the urban poor living in Old Fadama face repeated rounds of eviction and forced displacement. I also explore how urban land management in Accra is fueled by demolition and flood control logics that paradoxically redirect and reorient the focus and politics of environmental health in Agbogbloshie. I explore how e-waste workers experience "slow violence," a term used to describe "violence that occurs gradually and out of sight, a violence of delayed destruction that is dispersed across time and space, an attritional violence that is typically not viewed as violence at all" (Nixon 2011, 2), as well as experiences of "fast violence" in the form of acute bodily distress and vital socioeconomic displacement. But Agbogbloshie is not simply a precarious space of destruction and impossible living. It is also a place where e-waste workers sustain a cultural life struggle in Ghana's urban margins.

Chapter 4 introduces the ways in which e-pyropolitics are embodied by exploring the illness narratives and bodily distress experiences of several copper burners. In this chapter, I draw on ethnographic narratives to explore how Agbogbloshie workers narrate, understand, and refer to their own bodily distress to make sense of the toxic exposures and environmental health risks they face. In particular, I attend to how workers refer to not only "internal" conditions of toxicity (e.g., lung and heart pains) but also the more visible "exterior" forms of harm—burns and scars—to make sense of their own toxic corporality. I also highlight the ways in which polluted bodies become the site of public health science and the focus of laudable environmental health risk mitigation efforts that fall short of making work in Agbogbloshie less toxic. In this way I explore how these health interventions inadequately address the complexities of life and work in a scrap metal market and environment where vibrant bodies, toxins, and postcolonial economies intersect.

Chapter 5 engages a critical discussion of Agbogbloshie as a site of toxics photography and e-waste ruination visualization. I explore how e-waste images make meaning and shape imaginations of the e-waste problem in Ghana, focusing on how fiery e-waste imagery distorts and mystifies life in Ghana's e-wasteland. I shall argue that this e-pyropolitical visualization parallels other photographic interventions focusing on ruination, industrial decay, urban rust, and various forms of infrastructural rubble, abandonment, and destruction. The chapter interrogates the e-pyropolitical gaze that makes Agbogbloshie both site and *sight* of digital rubble, revaluation, and toxic colonialism. In response to this contentious visual economy of e-waste, I consider the possible role of participatory photography as an alternative technique of ethnographic e-waste visualization. I explore how worker-based forms of witnessing e-waste can help justify and provide a methodological grounding for the very decolonization of e-waste studies in Ghana in particular.

Chapter 6 engages the looming uncertainties of e-waste management and innovation in Ghana amidst serious infrastructural challenges, a hyper-debt crisis, and the COVID-19 pandemic. Agbogbloshie has become a neoliberal intervention environment where e-waste recycling innovation and economization are meshed with optimistic green developmentalist agendas, projects, and discourses of hope. As explored in this chapter, much is needed to address Ghana's e-waste challenges, but especially important are environmental health interventions and e-waste projects and partnerships directly informed by emerging "just transition" (White 2020) and decolonization debates. Finally, to conclude the book, it will be argued that while Agbogbloshie is certainly a toxic environment, this toxic slot[52] ought not be a trap for keeping alive or recycling the hypernegativity and dystopic gaze shaping much of Ghana's fraught e-wasteland narrative. In light of this challenge, the book ends with a call to resituate e-waste studies in relation to the uncertainties of "coronavirus capitalism" (Klein 2020b; see also Davis 2020; Garrett 2020) and the sparks of resistance, hope, and solidarity emerging from global Black Lives Matter movements and critical engagements with racialization, dispossession, extraction, and toxic Black ecologies (Hare 1970; Roane and Hosbey 2019; Hosbey, Lloréns, and Roane, forthcoming).

[52] This phrasing derives from the powerful insight of Trouillot (1991), who rightfully critiqued anthropology's distorted obsession with the "savage."

1

Amidst Global E-Waste Trades and Green Neoliberalization

Countries in the global north definitely dump e-waste in the global south—that is not under debate. What is under debate is what an exclusive focus on that particular aspect of spatial politics does to potential alternative circulations, politics, and actions.

—Max Liboiron (2014)

Material that can't be recovered goes to the fire.
—Lambert Faabeluon, Ghana Environmental Protection Agency

Without a doubt, the most common question I get when I explain to friends, family, and co-workers that I research the cultural, health, and environmental politics of e-waste in Ghana is, "How does all that waste get there?" My anticipation of this question is by now almost automatic, and I have to admit, it is an excellent, yet always challenging, question to comfortably respond to and navigate. One could begin to respond with a simple reference to the stance of a US Interagency Task Force on Electronics Stewardship, which admits there is little, if any, reliable data on exported e-waste (Interagency Task Force on Electronics Stewardship 2014, 14). But before even attempting a satisfactory response to this common question, we need to ask a deeper and more burning question: "Why has Africa in particular, and the south in general, come, in significant respects, to anticipate the unfolding history of the Global North? Why, for good or ill, are the material, political, social, and moral effects of the rise of neoliberalism most graphically evident there?" (Comaroff and Comaroff 2012, 13). This question gets at the heart of the political economy and ecology of e-waste in Africa in general and Ghana in particular. So, what really has caused Agbogbloshie to become a notable global hot spot of e-waste ruination? Why, for example, are mountains of junked photocopiers piling up there (Figure 1.1)?

Burning Matters. Peter C. Little, Oxford University Press. © Oxford University Press 2022.
DOI: 10.1093/oso/9780190934545.003.0002

Figure 1.1. Photocopier accumulation in Agbogbloshie. Photo by author.

What are the economic, political, social, and moral effects of this postcolonial material settlement? What forms of optimistic "environmental" intervention take root in Agbogbloshie, and how are these intervention projects informed, either directly or indirectly, by neoliberal patterns and shifts in global e-waste governance and advocacy politics?

Complex and uncertain global patterns of e-waste trades and "stewardship" initiatives and policies help explain how and why Agbogbloshie became a node in metal scrap economies. The truth is, within broader "geographies of waste" (Millington and Lawhon 2019) e-waste circulates globally with very limited stewardship and surveillance, leaving certain regions and communities of the Global South more vulnerable to highly toxic e-waste recycling practices. Unlike flows of e-waste found in the Global North, many wastes found in Accra have an international origin (Grant 2006; Grant and Oteng-Ababio 2016). This reality helps explain why and how the information technology (IT) revolution is changing established global e-waste trade dynamics. For example, instead of only the United States, Canada, and countries in Europe, now China, India, and South Africa are emerging as dominant producers, sources, and destinations for electronic discard, as well as significant players in the e-waste circular economy. According to a recent report, Australia, China, the European Union, Japan, the United States, and the Republic of Korea are the dominant producers of e-waste; but per capita (individual) production of e-waste between north and south tells a different story. For example, to understand per capita e-waste production, the United States and Canada dominate. In these states, per capita e-waste production is roughly 20 kilograms annually, while in the European Union the figure is approximately 17.7 kilograms. But in Africa, a continent with roughly 1.2 billion people, per capita e-waste production is just 1.9 kilograms annually (Baldé et al. 2017).

The Basel Convention, which went into effect in 1992, outlines international regulations for the trade of hazardous wastes in order to minimize exploitation effects from the waste trade that are felt by developing countries like Ghana. The convention's treaty explicitly contains policies on the restriction of transboundary movements of hazardous wastes except where it is perceived to be in accordance with the principles of "sufficient management" or where "electronic stewardship" (e-Stewards) principles are imposed. In short, this legislation leaves large gaps in understanding, fosters misinterpretation, and reinforces contentious loopholes as evidenced by the significant streams of toxic electronic discard in Ghana that in fact still originate from even signatory nations, mainly countries in the European Union. As Richard Grant, an e-waste geographer and internationally recognized expert on e-waste issues in Ghana, recently pointed out, "Europe is by far the most important exporter of used computers to Ghana, followed by the United States. Much of this trade is considered donations to accord with the Basel Convention (which regulates the transport of hazardous waste), but non-working devices are often included in exports" (Grant 2016, 25). Grant adds that "Flows into Ghana from Asia, the Middle East, and elsewhere in Africa are also rapidly increasing. Some of this regional traffic is European and North American traffic that is concealed by routing container traffic to Ghana via Hong Kong, Durban, Mombasa, and Dubai" (2016, 25). It is also suspected that negative media exposure in China and e-waste contraband reporting in Asia, in particular (Black 2004), have led to significant increases in Ghana's e-waste imports (Grant and Oteng-Ababio 2012).

In addition to exposing the global political economy and ecology of e-waste trades and policies in Ghana, this chapter provides a critique of a recent neoliberal shift in discourse and political actions taken by the Basel Action Network (BAN), the primary nongovernmental organization (NGO) involved in overseeing and regulating the international trade in hazardous wastes such as e-waste. By 2000, century (in 2000) BAN began framing the e-waste crisis in terms that emphasize the role of corporate social responsibility[1] and harnessing technical expertise to manage hazardous wastes in "green" and environmentally sound ways. As argued in this chapter, the assemblage of environmental interventions deployed in Agbogbloshie recycle common NGO interests and goals of turning places and spaces of extreme environmental suffering into opportunistic sites of eco-neoliberal experimentation.[2] NGO intervention in Agbogbloshie shadows other international e-waste advocacy approaches which tend to converge with the competitive economic discourse of the neoliberal state. With a critical eye on an international NGO fixed on global pollution control, I account for the ways in which Pure Earth (formerly known as the Blacksmith Institute) reinforces the neoliberal narrative of global e-waste toxics. Pure Earth operates as a "solutions-based" organization yet is also supported by a network of "certification actors [that] play critical roles in securing narratives like 'digital development' or 'toxic trade' as the hegemonic commonsense understanding of the e-waste problem" (Pickren 2014, 28). I argue that while on the surface it appears that Pure Earth emphasized an e-waste intervention that took seriously solutions to effectively reduce the environmental health risks of e-waste recycling labor, this neoliberal intervention followed market-driven disposal options that only benefited a small subset of workers in Agbogbloshie. The Pure Earth intervention has also created more conditions for confusion and precarity[3] than solutions and answers, a fact that is curiously avoided in the NGO's own discourse and reporting. Pure Earth is yet another green organization and agent within a larger, more systematic neoliberal environment of waste management and practice. While Ghana has been described as a key "neoliberal pacesetter" (Chalfin 2010, 29) in Africa, especially since its adoption of aggressive neoliberal reforms in the 1980s, more recent debates

---

[1] For an excellent global overview of ethnographic engagements with corporate social responsibility, see Dolan and Rajak (2016).

[2] As Michelle Murphy cleverly notes, "neoliberal experimentality" involves an "infrastructure of experiment" that is tightly bound to logics of a neoliberal state that sees "a world open to intervention and productive change" (2017, 90–91).

[3] Fixing precarious labor conditions wasn't the goal of this e-waste intervention. In fact, it is important to recognize that "the subject of precarity and precarious working conditions in advanced, post-industrial economies is often premised on the false binary of precarity–stability" (Ivancheva and Keating 2020, 251). Precarity, in this way, must be understood in relation to the "stable career" shift in the post–World War II years as professional middle classes and labor organizations in the Global North gained momentum and centering.

over Ghana's e-waste problem are showing signs of yet another iteration of neoliberal interventions on/in global e-waste recycling economies, policies, and laws.

Ghana in an E-Wasted World

Discarded electronics are one of the 21 most pressing global environmental problems of the twenty-first century (United Nations Environment Programme 2012). Consider what has happened in just a 4-year period. Between 2010 and 2014, the United Nations estimated that the amount of e-waste produced in the world each year increased from 33.8 metric tons to 41.8 metric tons. Also between 2010 and 2014, 3 million metric tons of waste from computer, cell phone, and other IT equipment waste was produced. At the level of individual consumption, the amount of e-waste generated per person is also on the upswing. For example, a 2017 report projected that rates would rise to 14.8 pounds per person per year in 2018, compared to known rates of 11 pounds per person per year in 2010 (Baldé et al. 2017). In 2014, the United States placed first as the world's leading e-waste producer, followed by China.[4] All of the major e-waste-producing countries, which in addition to the United States and China include Canada, South Korea, Japan, Australia, and several European countries, have over the years increased their exportation of waste electrical and electronic equipment (WEEE). This trend is due largely to the fact that both increased restrictions on e-waste disposal and rising recycling costs in these developed countries have become major drivers of increased WEEE exports. For example, a report by the International Labour Organization called *The Global Impact of E-Waste: Addressing the Challenge* reiterates earlier findings of the US Environmental Protection Agency, claiming "it was ten times cheaper to export e-waste to Asia than it was to process it in the United States. The incentives for e-waste movement, both legally and illegally, are thus enormous" (Lundgren 2012: 14). So, it is important to recognize that e-waste economies (both formal and informal) exist within a world economic and political system involving no shortage of extra-state "il/legality" (Nordstrom 2007).

---

[4] In fact, a report by the US International Trade Commission (2013) used the term "used electronic products," rather than e-waste. These materials, as noted in the report, are "classified as working electronic products and parts to be refurbished and resold, or as non-working goods to be recycled into scrap materials."

In 1991, just seven years before the BAN was launched, 25 African nations signed an e-waste treaty called the Bamako Convention. Like the BAN, this convention was designed as a treaty to reduce exploitative patterns of e-waste dumping in Africa. There was international optimism about this regionalized convention. Unfortunately, little implementation has taken place to date. In fact, as one official in Ghana's Ministry of Environment told me in an interview, "The treaty is very weak. It forgets about the economic nature of Africa. Africa and Ghana is a place of very porous borders. Ghana, like Africa itself, is about trade, trade, trade." The Basel Convention contains policies on banning the transboundary movements of hazardous wastes to developing nations that lack proper e-waste management systems, but, as Grant (2016) points out, the Bamako Convention's treaty builds on this by prohibiting the import of all hazardous and radioactive wastes into the African states that sign and ratify it. Ultimately, the purpose of this treaty is to minimize and control the flows of e-waste that travel within and between African states and to ensure that electronics are disposed of and recycled in an environmentally sound manner. But a primary contradiction of market-based environmental governance seems to overshadow both BAN and Bamako Convention logics. In other words, "in the competitive race to reform the market, certification actors engage in a kind of representational race-to-the-bottom in which the simplest ideas become embodied in the label and sold to consumers" (Pickren 2014, 40). In Ghana, as in many regions of the Global South, synthesizing economic development goals with international and regional e-waste governance ideals is not easy. Workers in Agbogbloshie engage in this toxic labor simply to make a living, even if this socioeconomic livelihood leads to their own embodied suffering (see Chapter 4). Basic everyday survival needs override worker compliance with BAN and Bamako Convention logics and restrictions. It could even be argued that this embodied noncompliance is actually what keeps copper scrap metal economies afloat in Ghana and beyond.

Copper scrap recycling is a booming sector of world metal economies. For one thing, growth in this market is due to the simple fact that recycling of copper from electronic discard is far cheaper than producing copper from ores in places like Chile, the world's leading copper producer. According to experts at the European Copper Institute, recycling copper can use up to 85% less energy than copper ore production. Ironically, using this energy-efficiency logic, the method of copper extraction and recycling practiced by "the burners" in Agbogbloshie makes sustainable energy sense, even if such a practice comes at the cost of extreme environmental contamination and toxic embodiment. What is for sure is that copper extractivism, recovery, and recycling are dynamics of the copper circular economy that, when viewed from the perspective of Ghana, reveal important context-specific trends. Ghana, for example, currently imports used electrical and electronic equipment from 147 countries. In 2004, Ghana started to formally import used electronics and electrical equipment, and by 2009 it is estimated that imports had reached 215,000 metric tons, with nearly 70% being classified as e-waste (Schluep et al. 2012). It is therefore critical not only to couch the e-waste problem in Ghana within a global system of e-waste and scrap metal trade framework favoring north-to-south dumping but also to consider the growth of these imports and the vibrancy of domestic flows of not just solid wastes (Baabereyir, Jewitt, and O'Hara 2012) but also flows of refurbished electronics. For example, the volume of refurbished electronic devices that are in circulation in Ghana far exceeds the tonnage of e-wastes being imported (Schluep et al. 2012).

Ghana's e-waste narrative is incomplete without careful attention to certain truths about domestic electronics consumption and Ghana's involvement in the global IT revolution. As Grant (2016, 23) notes, "mobile phone subscriptions in 2012 (per 100 people) surpassed the number in the United States. Scavengers operating at very high collection rates for electronic devices enable the urban mine to function." In other words, rising imports and sustained growth in domestic collections *combine* to further explain why e-waste accumulates at Agbogbloshie. It also changes the blame game and finger pointing that consume many debates about e-waste dumping in Africa and other regions of the Global South. This rethinking of e-waste in Ghana corresponds with the ethical shift occurring in critical e-waste studies, which asks "Is a system of 'ethical' or 'fair' trade in e-waste . . . a viable alternative to the existing strategies of national and international legal prohibitions against e-waste exports from 'developed' to 'developing' countries?" (Lepawsky n.d.). This question calls attention to a deeper discussion of the *political* contextualization of global waste and environmental justice politics writ large.

Shifting Global Waste Politics

Many e-waste disputes and conflicts, as well as urban waste struggles in general, are symbolic of other global crisis discourses signaling the waning capacity of state and international institutions to effectively manage and control techno-corporate power and capitalism. For this reason, recent environmental governance studies point to the growing need to more directly engage the "political-*economic* opportunity structure" (Pellow 2001, emphasis in original), the enduring power of profit (Gareau 2013), and the actual limitations of civil society influence on environmental governance itself (Gareau 2012). In a case study of a "good-neighbor" agreement, Pellow (2001, 54) explains that with the declining power of nation-state governance, "many arms of the U.S. environmental movement have pursued a combination of confrontational and accommodationist strategies that simultaneously challenge and accept the basic goals of the corporation: profit." In other words, some social movements reason that if corporations are the ones pulling the strings, then activists must, like corporations, play the market-based game. This is what actually fueled most corporate social responsibility campaigns emerging in the 1990s: movements to moralize, humanize, and ecologize business ethics to "[leverage] the power of the brand to mobilize consumer–corporate–NGO 'partnerships' for development ends" (Dolan and Rajak 2016, 7).

Critical perspectives on waste and pollution helped spark early engagements with the dynamic concept of environmental justice (Bullard 1990; Mohai and Bryant 1992; Szasz 1994). With corporate power accelerating in the twenty-first century, these effects have taken on a more intrusive global dimension, with scholars focusing on international environmental justice issues through the direct and indirect trade in hazardous wastes (Clapp 2001; Faber 2008; Pellow 2007). In the absence of regulatory oversight and a decrease in transportation costs, toxic industries began searching beyond national borders for cheaper sites to dispose of these wastes. Regions of the Global South, especially states in Africa and Asia, became susceptible dumping grounds for toxic discard (Faber 2008) and are now central zones of the broader "wasteocene" (Armiero 2021). Ironically, even though during the 1980s and 1990s a cornerstone success for the environmental justice movement in the United States was direct action against polluting industries, what followed was actually the increased offshoring of unwanted material hazards, including byproducts of production and postconsumer wastes like e-waste. This is one reason e-waste politics in Africa, including Agbogbloshie, even began to make global waste headlines.

Media coverage of e-waste issues didn't really gain any momentum until around 2004. In fact, between 1980 and 2012, major global news outlets released roughly 1,000 stories on e-waste, but only eight of them dated before

2002. This broadly suggests that, although information on the e-waste trade has been public at least since the 1990s, it has only gathered significant US media attention more recently, with attention to the issue growing alongside the shift in the activists' discourse (Little and Lucier 2017). Additionally, a few highly publicized incidents of transnational waste shipments gone awry were actively and creatively resisted by transnational environmental organizations such as Greenpeace. An illustrative example is the *Khian Sea*, a ship carrying toxic fly ash[5] from a Philadelphia incinerator. Early in the ill-fated journey of this ship, the crew deliberately mislabeled some of the toxic cargo as fertilizer and dumped it on a Haitian beach at night after being refused entry into the country. Activists from Greenpeace, as well as locals from Philadelphia and Haiti, launched "Project Return to Sender" in response. A sample of the tactics employed in this campaign included shipping hundreds of envelopes with samples of the toxic ash to the mayor of Philadelphia with "Return to Sender" in big letters, as well as distributing "wanted" posters around Philadelphia with the mayor's picture and the text "Wanted: for Environmental Racism" (Pellow 2007). After a number of years, a portion of the waste was returned to US soil. Campaign narratives like these are often repeated by scholars and activists tracing the inception of the Basel Convention on the Transboundary Movement of Hazardous Waste, which had its first meeting in 1989 and entered into force in 1992. This is the first global convention that regulates the movements of hazardous waste across national borders, and it has spawned some regional versions such as the Bamako Convention in Africa.[6] Most scholarly accounts of the Basel Convention highlight the unusually strong presence of the environmental NGO Greenpeace (Clapp 2001; Pellow 2007). Not only is Greenpeace consistently cited as the primary source of data on the extent of the global trade in hazardous wastes in the years leading up to the convention but it is also credited with forging enough unity between the G-77 (less developed) countries for them to prevail on a consensus vote to adopt the so-called Basel ban in 1994.

This ban would halt the transfer of hazardous wastes destined for both disposal and recycling from wealthy countries (e.g., Organisation for Economic

---

[5] Coal ash is the waste that is left after coal is combusted (or burned) and includes fly ash. "Fly ash" is the term for the fine powdery particles that are carried up the smoke stack and captured by pollution control devices. Most coal ash comes from coal-fired electric power plants.

[6] Enacted in 1998, the impetus for the Bamako Convention arose also from the failure of the Basel Convention to prohibit trade of hazardous waste to less developed countries and the realization that many developed nations were exporting toxic wastes to Africa (Bamako Convention 1991).

Co-operation and Development [OECD] countries and Lichtenstein) to less developed countries.

One of the leaders of the Greenpeace effort in the Basel Convention was Jim Puckett, who later formed a separate environmental NGO, the BAN. BAN's most concrete objective has been to bring Basel's ban amendment into force, but more generally its objective, as stated on its website, is "to champion global environmental health and justice by ending toxic trade, catalyzing a toxics-free future, and campaigning for everyone's right to a clean environment" (BAN n.d.).

Until recently, its actions to this end have ranged from direct, on-the-ground investigations of toxic dumping and recycling sites around the globe (aimed at providing information as well as raising awareness), working directly within the political structure at meetings of delegates of the Basel Convention, to practicing what they call "coalition campaigning," where they join up with local and transnational NGOs in order to raise awareness and empower others to act in their local contexts.

BAN's focus on the growing e-waste problem was made public beginning with a 2001 investigative "mission" to Guiyu, China. Here, the activists visually and empirically documented the ecological and health horrors associated with the informal economy that arose around e-waste "recycling." Since then, they have made four other trips to central e-waste villages in China, Ghana, and Nigeria. At the same time, major news outlets in the United States have taken an interest in the global e-waste crisis, with BAN's images and exposed e-waste recycling sites featured in mainstream outlets such as PBS's *Frontline*, CBS's *60 Minutes*, and the *New York Times*. In an interview, one representative of a European government think tank pointed out that BAN's ability to attract the media comes down to the fact that "BAN wants to reduce every issue to a one sentence sound bite." For sure, BAN faces a particular challenge of having to target its efforts squarely on the United States, the world's largest volume generator and exporter of e-waste. This is a practical challenge because the United States has not even become a party to the Basel Convention (the United States along with Haiti and Afghanistan are the only countries to have signed but not ratified the convention), and the present laws under the Resource Conservation and Recovery Act do not ban the export of electronic waste to non-OECD countries. These are just some of the "loopholes" conditioning the direction and character of global e-waste flows (Kirby and Lora-Wainwright 2015). E-waste recyclers in China, for example, admittedly struggle to meet BAN-proposed standards (Little and Lucier 2017). For example, one e-waste expert interviewed worked with the United Nations Environment Programme and openly explained that "It could be bothersome, since they want a higher than reasonable level of environmental purity. Guidelines could form a more informed regulatory system. The obstacle is that it is always cheaper to not follow some of these practices. Developing countries are so unregulated." While an "unregulated" environment may be a more favorable option for capital-driven actors in the global e-waste trade, this point also highlights deeper spatial logics of environmental racism which have historically inspired research on environmental justice advocacy and anti-racism activism (Pellow 2007; Checker 2005; Andrews 2016).[7]

BAN has long described the waste trade as something akin to garbage imperialism and an extreme case of unequal global environmental governance (Okereke 2008), even a process linked to "toxic colonialism" (Akese 2014). All these discourses are logically congruent with the political changes that BAN has traditionally pushed for, particularly the global north–south waste trade ban. However, in the process of developing a voluntary third-party e-waste recycling standard in the United States, this discourse of *exploitation* was sidelined in favor of a discourse focused on technical *expertise* in e-waste management (Little and Lucier 2017), a theme that steers current efforts in Ghana to manage its e-waste mess (see Chapter 6). BAN, it turns out, appears to be moving farther away from advocacy approaches couched in environmental justice and anti-racism, away from an approach that more directly "challenges the status quo rather than fixing or tinkering with a system grounded in domination, racial terror, and colonial control" (Sze 2020, 14). This has resulted in a BAN paradox. For example, in an early 2003 draft document on the formation of Basel-sponsored public–private partnerships, BAN stressed the need to "find and develop practical, sustainable solutions to delink economic development and the waste it traditionally generates." Later, in 2004, BAN called for a shift in focus on health and pollution to more "fully comprehend the repercussions of waste trade and its impact on the local environment and community. This is the other side of the economic coin of the trade in hazardous wastes that is consistently unaccounted for" (Basel Action Network 2004).

There is another dimension to BAN's discursive shift, specifically the gradual use of the word "primitive" to describe the people and processes

---

[7] These spatial logics have also been a subject of focus for global studies of the Black Lives Matter movement. In fact, in 2016, the *World Policy Journal* published a special issue (volume 33, issue 1) on "Black Lives Matter Everywhere."

involved in the informal e-waste recycling in non-OECD countries like Ghana. Beginning in 2003, the use of "primitive" is apparent both in interviews with and statements made by BAN and in official convention documents submitted by BAN and Greenpeace. The deployment of the term appears in a variety of contexts, from Greenpeace's report on shipbreaking in India in 2003 to interviews with Jim Puckett on National Public Radio and for *Frontline* to Puckett's own editorial in the trade journal *E-Scrap News*. Given the sensitivity to neocolonial dynamics that is implied in BAN's earlier assessments of the waste trade, this choice seems especially discordant with BAN's earlier language. The implication of this shift, from populations working in the informal economy previously being described as "exploited" to their being "primitive," seems to downplay the inequities of neoliberal capitalism. Instead, the purported e-waste "crisis" is due to a lack of proper technical expertise to manage these wastes in a socioeconomic environment that is "underdeveloped." This discourse can be contrasted with the discourse of "toxic colonialism," as well as other instances where BAN characterizes e-waste recycling as an inherently dangerous process, regardless of the degree of technical expertise, technology, and processing infrastructure. In sum, BAN has largely succeeded in increasing the level of public interest in the e-waste problem through embracing a discourse of economic logic and technical expertise, but this has resulted in the unfortunate sacrifice of more righteous social critiques of exploitation, environmental injustice, and toxic colonialism.

Another thing to consider is that even "green" certified e-waste recycling practices in the Global North aren't free of these neoliberal capitalist logics or devoid of serious occupational hazards. For example, there are several major instances where the e-Stewards standard has watered down or even reversed the previous policy positions taken by BAN, the International Campaign for Responsible Technology, or even the original e-Stewards proclamations. The first major concession concerns the standard for a corporation being labelled an "e-Stewards enterprise." So far, major corporations such as Bank of America, Alcoa, Wells Fargo, Samsung, and Capital One have been labeled e-Stewards enterprises, enabling them to use the e-Stewards logo on their materials and enabling e-Stewards to place their corporate logo on the e-Stewards materials. As noted in Little and Lucier (2017), electronics refurbishing industry representatives highlight both benefits and continued risks of even certified e-Stewards contracts. According to one of these representatives, "Certification is a 'de facto' requirement now. This has caused major changes in the industry. Even following certification there are still risks." This informant added that the National Institutes of Occupational Health and Safety (NIOSH) has, for example, found elevated lead levels even in certified e-waste recycling facilities. Currently, NIOSH is conducting epidemiological studies of electronics recyclers in the United States; and initial indications suggest that, even in e-Stewards-certified facilities, workers are facing risks of toxic exposures, and the employees in these facilities are often "people of color" and are paid low wages, raising further concerns about new cases of environmental injustice for e-waste recycling workers.

Neoliberal Intervention and Decomposing "Development"

In his ongoing anthropological study of neoliberal intrusion and transformation in Africa, anthropologist James Ferguson (2006) reminds us that many anthropological critiques of development in Africa can be understood as studies of "modernity's decomposition." Amidst debates over the question of modernity in Africa in general, he urges us to take caution when confronted with development discourses that treat Africa as an "alternative" place and space of political-economic development projects:

> [T]he application of a language of alternative modernities to the most impoverished regions of the globe risks becoming a way of . . . avoiding the question of rapidly worsening global inequality and its consequences. Forcing the question of Africa's political-economic crisis to the center of the contemporary discussion of modernity is a way of insisting on shifting the topic toward . . . the enduring axis of hierarchy, exclusion, and abjection, and the pressing political struggle for recognition and membership in the emerging social reality we call "the global." (Ferguson 2006, 192–193)

A similar approach can be taken when confronting the contemporary discussion of Ghana achieving recognition as an experimental space and market for alternative e-waste recycling in Africa.

In November 2013, Pure Earth (formerly the Blacksmith Institute) and Green Advocacy Ghana began a pilot project to combat Ghana' e-waste problem. This was the first of the so-called green neoliberal alternatives to take root. They started by setting up what they called "a basic e-waste recycling facility that would enable recyclers to stop burning wire and instead strip it in a way that was efficient and profitable" (Pure Earth 2015). Funding for the pilot project came from the United Nations Industrial Development Organization (UNIDO), through the Global Alliance for Health and Pollution (GAHP). The GAHP formed in 2012 to address "pollution and health at a global scale" and is an alliance of Pure Earth, the World Bank, the United Nations Environment Programme, the United Nations Development Programme, UNIDO, the Asian Development Bank, the European Commission, and numerous ministries of environment and health from many low- and middle-income countries (Figure 1.2).

The project in Agbogbloshie began with a series of open community forums that brought together important stakeholders including government officials, local businesses, the Greater Accra Scrap Dealers Association (GASDA), and others working outside the scrap metal recycling market. Through these forums, project plans were discussed and community members were able to voice their opinions, questions, and concerns. According to Pure Earth, these community forums "vastly improved relations between the recyclers and the government of Ghana," adding that

Figure 1.2. Original e-waste recycling project banner. Photo by author.

Figure 1.3. The granulators in action. Photo by author.

"Since the recyclers are participating in an informal sector activity, the government of Ghana had previously been wary of GASDA." As a result, Pure Earth worked closely with GreenAd Ghana to find ways to "improve relationships and show government officials the economic value of the emerging e-waste recycling sector in Ghana." The National Youth Authority (NYA),[8] the state agency that owns the land where the Agbogbloshie scrap market sits, agreed to the pilot project and dedicated a piece of land specifically for the new recycling facility. The facility was created using three 40-foot intermodal containers (standard shipping containers) and houses four mechanized wire-strippers, two for small-diameter wires and two for large-diameter wires (see Figure 1.3). While the NYA has given permission to build the facility and is in full support of the project, the scrap metal market is part of a larger food and goods market. Knowing that the e-waste recycling sector

---

[8] According to the NYA's website, this government agency "exists to provide relevant and conducive environment that defines and supports the implementation of effective frontline youth empowerment practices, focusing on young people's participation in socio-economic and political development [whilst] facilitating private and third sector provider investments in youth empowerment" (www.nya.gov.gh).

could expand, requiring additional space for operations, project managers decided to use these containers since they can be transported easily.

Most importantly, the pilot project, according to Pure Earth, has demonstrated the economic viability of switching from burning to stripping of copper wires. The pilot project was deemed "completed" in September of 2014 with the official opening of the e-waste recycling facility, an effort heralded as a solution to modernize, mechanize, and better manage Ghana's e-waste recycling work. According to Pure Earth, the project's success called for changing e-waste worker behavior: "Behaviour change is an on-going process: by first piloting a small number of machines, the recyclers and GASDA are beginning to understand the benefits of using these tools instead of burning . . . and to better understand the workloads and needs of the recyclers" (Pure Earth 2015). The facility is managed by the Agbogbloshie Scrap Dealers Cooperative and is jointly owned by GASDA, GreenAd, and the NYA. Members of these organizations formed a management committee, which also included an environmental health and safety officer and an accountant, with oversight by the Ghana Environmental Protection Agency. Six workers in the scrap yard were trained on the use, maintenance, and repair of the stripping machines.

The intended goal of the project was to get wire collectors in the scrap market to sell their wires directly to GASDA and the facility. The wires would then be stripped by the on-staff recyclers trained to use the machines, and then GASDA would sell the copper. Pure Earth claims that "an estimated 450 pounds of copper are recycled at the facility every month, with the recent addition of approximately 40 pounds per month of aluminum also now being recycled" (Pure Earth 2015). It is well understood that this is simply a "pilot" project, a starting point in Ghana's effort to modernize its e-waste recycling sector. Managers of the project learned early on that workers collect and burn a lot of small-diameter wires, so the purchase of machines that could strip small wires was a needed investment. Another lesson learned was that instead of importing stripping machines from China, machines could be found within Ghana, further supporting local businesses. Ultimately, GASDA had a vision to promote Agbogbloshie as a recycling knowledge center by setting up a model e-scrap facility that would "protect livelihoods while minimizing the adverse health and environment risks of scavenging and exposure to toxic substances."

As a result of the pilot project, Pure Earth was able to expand the project with funding from a major petroleum industry partner and investor, the

Addax & Oryx Group Limited, a company involved in capital investments in energy, real estate, and a major global provider of bitumen, bunkering, and other oil and gas services. Additionally, the company engages in oil and gas exploration and production across Africa and the Middle East.[9] This additional corporate funding allowed for several new machines to be purchased for the recycling facility, including a granulator and separator, which would allow for small copper and aluminum wires to be "properly" recycled. Additionally, Pure Earth and its financial partners provided "formal accounting and business training" for senior members of GASDA. It was assumed that these trainings would help continue to build GASDA's capacity to run a recycling facility. Additionally, with the extension of the project, Ghana Health Services pledged to provide "health and safety trainings" for workers both within the new facility and throughout the market. Amidst these various Pure Earth interventions, the underlying message that was spread throughout Agbogbloshie was that "Burning Is Bad" (*kona ne bad*) (see Figure 1.4).

In the midst of Ghana's presidential elections in 2016–2017, a race won by Nana Akufo-Addo,[10] plans for a large urban e-waste facility in Agbogbloshie were under development. The project was a collaborative effort between Ghana's Ministry of Environment, Science Technology and Innovation and the Environmental Protection Agency. According to one local news source, it was reported that with the construction of the new facility, Ghana would save the roughly $200 million it spends annually on "containing the health impacts of electronic waste disposal." The facility, it was suggested, would also help restore the "once serene ecological zone" of the Korle Lagoon, commonly described as "one of the most polluted bodies of water on the planet" (Boadi and Kuitunen 2002). The ultimate goal of the facility is "to re-strategize to deal with the adverse health impacts associated with the menace of electronic waste." Yet health and environmental concerns were not the only issues fueling this multimillion-dollar effort of restrategization. A project consultant working on the proposed facility, who also serves as the

---

[9] According to the United Nations most recent *World Investment Report* (United Nations 2019), foreign direct investment in Ghana is mostly focused on gas exploration and production and gold mining. For example, the Eni Group is currently the largest investment firm supporting the expansion of the Sankofa gas fields, and investments in gold mining are dominated by Asanko Gold Ghana Ltd., which is now partially owned by the South African company Gold Fields Ltd.

[10] Taking office in January 2017, Nana Addo Dankwa Akufo-Addo previously served as Ghana's attorney general from 2001 to 2003 and as minister of foreign affairs from 2003 to 2007 under the Kufuor administration.

Figure 1.4. *Kona Ne Bad*, or "Burning Is Bad."
Source: PureEarth.org

business development manager for SGS West Africa, added "From my economic point of view, we will be saving the country on the average about
$300 million by establishing this facility which is now going to engage in a
comprehensive value chain recycling for Ghanaians. The benefits I must say
are unquantifiable."

While on the surface the planning of this new facility seems to be the
brainchild of Ghanaian officials, the optimism fueling strategies to make
Ghana home to a fully "modern" and "certified" e-waste recycling facility indirectly stems from the e-sustainability ethos propelled by the international
e-waste recyclers' "Pledge of True Stewardship." The original version of this
pledge was drafted in 2003 and contains the following provision: "VIII: We
agree to support Extended Producer Responsibility (EPR)[11] programs and/
or legislation in order to develop viable financing mechanisms for end-of-life
that provides that all legitimate electronic recycling companies have a stake
in the process" (Smith, Sonnenfeld, and Pellow 2006, 308). However, by the

---

[11] As Lepawsky (n.d.) notes, "EPR is intended to internalize the costs of product disposal to
manufacturers and thus incentive them to modify their products to be more durable, repairable, and/
or recyclable. Typically, however, EPR for electronics is instantiated as a form of extended consumer
responsibility by passing costs for eventual recycling to purchasers of new electronic equipment." For
further details on EPR and electronics life cycle assessments, see reports by the Institute for Electrical
and Electronic Engineering.

time the e-Stewards standard was rolled out in 2010, this provision of the pledge was no longer included.

Currently, the e-Stewards policy is to prohibit exports of hazardous e-wastes from developed to developing countries, even if they work to certify facilities like the one proposed for Agbogbloshie or similar facilities being designed in other developing countries. At the same time, a recent announcement on the e-Stewards website states, "With the help of our generous sponsors, we were able to have a presence throughout E-Scrap [an annual industry trade conference held in September 2015]. Jim Puckett, Executive Director of the BAN, was on a panel discussing exportation, which the e-Stewards Standard supports in certain situations." These "certain situations" where exportation is "supported" are those situations that are now considered legal under the Basel Convention. Moreover, with recent reinterpretations in the Basel Convention on the meaning of "waste," permissible exports now include virtually any form of e-waste that is being exported by the original manufacturers of the equipment, as is often the case for mandatory or voluntary producer take-back programs. In other words, if corporation X in the United States sends its obsolete Dell computers back to Dell for recycling, Dell can legally export those computers to a facility in China or Africa for recycling. E-Stewards has adjusted its position to support this change in Basel policy (even though the BAN activists themselves vigorously opposed this change at Conference of the Parties 11 in Geneva), provided that the recycling facilities are e-Stewards-certified. Yet even these certified facilities are not considered free of occupational and environmental health hazards.

Witnessing the increase in global "certification" and e-waste recycling interventions and optimistic standardization gaining traction in Agbogbloshie in particular hints at neoliberal economization logics that reinforce a key logic of even "green" shifts in high-tech capitalism: "Economization is productive: it involves the organization of connectivity rather than disintegration" (Konings 2015, 2). It involves a multiplicity of new public–private partnerships and neoliberal "dreamworlds" (Davis and Monk 2008) informed by logics and transformations of e-waste policy and management that actively undermine and even undo righteous environmental justice critiques (Little and Lucier 2017). As Brown (2015, 17, her emphasis) puts it, "Neoliberal reason, ubiquitous today in statecraft and the workplace, in jurisprudence, education, culture, and a vast range of quotidian activity, is converting the distinctly *political* character, meaning, and operation of democracy's constituent elements into *eco-nomic* ones." This same neoliberal reasoning taints e-waste management and governance in Ghana, resulting in already marginalized workers being further devalued by e-waste economization efforts guided by the spirit of economic growth.

I began this chapter with a tough question that figures centrally in any discussion of waste management and e-waste problem-solving in Africa: "Why has Africa in particular, and the south in general, come, in significant respects, to anticipate the unfolding history of the Global North? Why, for good or ill, are the material, political, social, and moral effects of the rise of neoliberalism most graphically evident there?" (Comaroff and Comaroff 2012: 13). The recent optimism inspiring plans to build a modern e-waste recycling facility in Agbogbloshie extends the logics of neoliberal "development" at the same time that it aims to extinguish the toxic ills of wire burning, an established "primitive" e-waste recycling practice. This is an underlying imperative of Pure Earth and the Ghanaian state. But these e-waste interventions are part and parcel to ongoing neoliberal maneuvers in Ghana writ large and toxic statecraft practices across the Global South that increasingly expose how in fact "neoliberalism is built upon the never fully realized imperatives of imperial and developmental state building" (Chalfin 2010, 48). What is realized in Agbogbloshie are, at best, neoliberal interventions that seem to have all the failures and corruptions that mark the late and postcolonial African landscape (Mamdani 1996; Olivier de Sarda 1995). As some have noted, many of these intervention failures are the result of a general "misguided diagnosis" (Burrell 2016, 8) of how the e-waste "problem" in Africa is approached by organizations like BAN, Interpol, and the United Nations Environment Programme. These organizations tend to focus their attention almost entirely on regulating e-waste exports from the Global North. These reform efforts are certainly important, but they often overlook an important socio-technological fact: "The problem is that electronics (new and used) are being consumed in Ghana in greater and greater quantities. The work that needs to be done to handle the waste and protect these youth is work that must be done in country, in Ghana" (Burrell 2016, 8). Furthermore, alongside any regulatory discussion of global e-waste flows, trades, and neoliberal management politics in Ghana are the actual workers themselves, including many Dagomba laborers who struggle to sustain livelihoods in the bustling city of Accra. As argued in the following chapter, fiery e-waste accounts of life and labor in Agbogbloshie are hollow and misguided without acknowledging the ways in which migrant labor and urban–rural relations in Ghana, themselves relations symbolic of a Global Africa containing "contemporary circulations" of precious metals (Hodgson and Byfield 2017, 1), are also emerging in a sub-Saharan Africa dealing with significant climate disturbances.

2

"We Are All North Here"

Dagomba Labor Migrations and Meanings

*The body is made first and foremost to move, to walk, which is why
every subject is a wandering subject. . . . What is important is where
one ends up, the road traveled to get there, the series of experiences in
which one is an actor and to which one is a witness, and, above all, the
role played by the unexpected and the unforeseen.*
—Achille Mbembe (2017, 144)

Metal is where the real action is in electronics recycling.
—Grossman (2006, 227)

In Agbogbloshie, many paths are crossed. From Frafra-speaking laborers
from Burkina Faso selling goats to supply the local meat market to truck
drivers from Niger hauling onions to sell alongside Abose Okei Road to
Ghanaians from the country's Northern Region coming to Agbogbloshie
to ignite e-waste to extract and sell copper and aluminum to Nigerian businessmen at the top of Ghana's scrap metal sector hierarchy; most everyone in
Agbogbloshie, it seems, uses this marketplace as a *platform* of capital extraction to support life.[1] It is where mobility is temporarily localized and medi-
ated, serving as an urban scrap market with not only rural linkages but also
a place with sociopolitical connections informed by urbanized chieftaincy
relations. As I explore here, attending to contemporary conditions and lived
experiences of e-waste labor migration in Ghana is one way to go beyond
the *waste* focus of e-waste studies in Ghana. Instead, it will be argued, these
migrations exist along "circuits of extraction" (Arboleda 2020) and allow

---

[1] My use of "platform" here is informed by analyses and critiques of "platform capitalism" (Srnicek
2016), more generally, and platforms as "infrastructures on fire" (Edwards 2021), in particular.

Burning Matters. Peter C. Little, Oxford University Press. © Oxford University Press 2022.
DOI: 10.1093/oso/9780190934545.003.0003

us to think ethnographically about how, why, and where social and economic relations and narratives figure in the broader e-wastelanding story of
Agbogbloshie. This place demands a narrative pluralism that accounts for the
multiple stories and complicated lives passing through Agbogbloshie, a place
commonly told through a fixed "toxic" and pyropolitical lens emphasizing a
situated place of extreme pollution and global e-waste disaster. In this way,
Agbogbloshie is a situated urban scrapyard with deep and enduring hinterland links that call into question the usual images of and reductive narratives
about copper wire burners living on the front lines of toxic scrap labor. This
makes sense and is what these workers do, but it says little about the path they
are on and how they are leading complex migratory lives with weighty urban
and rural socioeconomic obligations to fulfill. As e-waste policy experts
note, a prerequisite of any efficacious e-waste intervention is understanding
ground-level social labor conditions (Lundgren 2012).

In this chapter, the focus is on the life experiences of copper wire burners
working in "urban" Agbogbloshie who originate from and sustain social ties in
the "rural" village of Savelugu. In particular, I focus on the story of one e-waste
migrant laborer named Ibrahim Akarima, who, like many other domestic[2]
e-waste migrant laborers, not only *ends up* in Agbogbloshie to burn electrical
and electronic discard to salvage valuable copper but also moves several times
a year between Ghana's Northern Region and the bustling coastal capital of
Accra. Understanding what goes on in Agbogbloshie, as one senior worker in
Agbogbloshie told me during the end of my first visit in 2015, requires paying
attention to worker connections to northern Ghana. In his words, "You go
north to understand what is going on here. We are all north here. To understand you must go there to see what we are doing here. That you must do.
I come here to go back north." During my first visit to Agbogbloshie in 2015,
it became clear that my e-waste ethnography needed to attend to how Ibrahim
and other workers navigate social relations, especially chiefdom obligations in
Savelugu, and toxic labor in Agbogbloshie. It became apparent that refocusing

---

[2] As Chalfin (2010, 201) rightfully notes, the out-migration of Ghanaians beyond its borders is
informed by significant forms of exodus in direct response to neoliberalism, which has had pan-
African effects:

> Although Ghana is touted as a model student of neoliberal reform, the country's sup-
> posed economic revival has stimulated and is fully dependent on an economics of exodus.
> After two decades of shrinking state support and the promotion of international trade at
> the expense of self-sustaining production (Aryeetey and Tarp 2000), a substantial per-
> centage of the Ghana populace is compelled to seek resources outside the country (Fine
> and Boateng 2000; Van Dijk 2003), illustrating what has become a pan-African dynamic
> (Kane 2002; Koser 2003; MacGaffey and Bazanguissa-Ganga 2000).

on city–hinterland relations helps resituate Ghana's "urban" political ecology of e-waste by further acknowledging the actual migrations of Agbogbloshie's workers, including their complex sense of belonging.[3]

Following the scholarly path blazed by James Ferguson in his *Expectations of Modernity: Myths and Meanings of Urban Life on the Zambian Copperbelt*, I am concerned here with "how we might understand the contemporary array of strategies through which workers manage the question of urban residence and rural attachment" (1999, 40–41) but also how Ibrahim, and other migrant e-waste laborers, manage urban scrap metal attachment in relation to enduring kin and social networks in Ghana's northern hinterland. E-waste labor migrations between Agbogbloshie and Savelugu form a kind of internal diaspora that has shaped north–south relations within Ghana itself. A dominant ethnic group and labor force in Agbogbloshie and Old Fadama, many Dagomba "form and maintain among themselves networks of sociality and community, feel connected to the homeland in the North. . . . Their diaspora is both a coherent social unit by itself and one node in a network of obligation and responsibility that ties it to the North" (Pellow 2011, 134). While the workers I have interviewed in Agbogbloshie never explicitly self-identified as living a diasporic life, their path to and from the north does involve a cultural diaspora that is now touched and informed by climate change. Making sense of these labor migrations now requires understanding agro-climatic changes in the Northern Region, especially environmental changes due to rising surface temperatures and declining rainfall (Tahiru 2019), which directly impact farming in the Northern Region. As explored here, migration narratives expose the ways in which the topic of e-waste labor migration in Ghana is in fact "a burning issue today" (Brosius 2018, 166), a topic with a pyropolitical twist that is linked directly to climatic rifts resulting from a planet that is physically "on fire" (Klein 2019).

Domestic Labor Migrations

Stemming from colonial histories of migration, natural resource conflict, and the adoption of neoliberal policies in the 1980s,[4] most migrants coming

---

[3] Migrant laborers, as Geschiere and Gugler (1998) note, also seek alternative forms of belonging and identification.

[4] As Pellow (2011, 136) points out, "Structural adjustment programs that were initiated in the 1980s only aggravated the disproportionate development. This was true throughout much of West Africa, as governments were forced 'to agree to restructure their economies in exchange for foreign

to Accra are from the Volta and Northern Regions of Ghana. Internal migration control was a primary strategy deployed in colonial Ghana, especially when it came to the shift to maximize production of export crops like cocoa and minerals like gold.[5] This production and extraction boom called for laborers from all corners of Ghana but most especially from what was called the "Northern Territories." During the British colonial era (1873–1957), much of this labor extraction was informed by violent policies aimed at forcing African bodies, through twisted logics of "the anatomy of power" (Butchart 1998), to work in the mineral mines and plantations along the Gold Coast Colony, what is present-day Ghana. The colonial state passed what was known as the Poll Tax Ordinance, an ordinance that forced subsistence farmers in the Northern Region to make impossible payments, leaving them no choice but to send members of their families to the southern coast to engage in cocoa farming for wages (Hill 1956).

The southern extraction of northern labor has a deep history in Ghana (Austin 1987) and is therefore certainly not unique to the socioeconomic and labor migration situation found in Agbogbloshie and Old Fadama. In fact, while labor extraction radically changed during Ghana's colonial period and with transatlantic slaving, "the dichotomy between the north and the south had already been laid over the preceding centuries and through the presence and predation of the Europeans and their forty-plus slaving ports, which still dot the Ghanaian shoreline" (Konadu and Campbell 2016a, 10). The Northern Region, for example, has long been severely deprived of vital infrastructural[6] investments (e.g., health, education, industrial agriculture, etc.), a situation that eventually led to increased out-migration of labor in the region. This pattern of out-migration is an issue that continues in present-day Ghana and figures centrally in the story and lived experience of migrant laborers circulating in and out of Agbogbloshie to extract what valuable metals they can from e-scraps. Even today the north–south split is a spatial arrangement that remains in Ghana.

---

loans. State budgets were cut, markets deregulated, public services privatized and state-owned enterprises and assets transferred into private hands' (Berry 2009, 25)."

[5] For a more recent ethnographic study of Ghana's gold mining industry, see Rosen (2020).

[6] As Appel, Anand, and Gupta (2018) recently put it, "Because infrastructures distribute vital resources people need to live—energy, water, information, food—they often become sites for active negotiations between state agencies and the populations they unevenly govern" (Appel, Anand, and Gupta 2018, 20–21).

As Dickson (1975) notes, coastal (southern) regions with exportable goods (i.e., cocoa, timber, and gold) generally have easy access to seaports and other supporting infrastructures for trade, a situation that continually reinforces north–south demarcations and inequalities. These north–south inequities are sustained and further exacerbated by national industrial economic policies. According to Awumbila, Teye, and Taru (2017, 986),

> Post-independence import-substitution industrialisation policies and more recently economic reforms and market liberalisation policies of structural adjustment policies since the early 1980s have done little to reduce these spatial inequalities and indeed have sustained the north to south migration flows [Songsore, 2009]. This situation has led to very high levels of deprivation in northern Ghana and in rural areas, with all major socio-economic indicators showing clear north–south, and . . . rural–urban differentials.

Rural-to-urban migration in Ghana has further intensified the growth of urban slums, where approximately 5 million people, or nearly 20% of the total population, currently live. The Ghanaian government expects this to increase by about 1.8% per year. In Accra alone, city authorities have identified about 78 slums and informal settlements of varying size and quality. Old Fadama is the largest of these urban slums and has some of the poorest living standards (Stacey and Lund 2016, 593), along with risks from disease, flooding, and flames, as in the case of a deadly 2012 fire (Owusu 2013).

Domestic migrations and structural displacements in Ghana have also been radically informed by natural resource extraction and energy development projects in various regions. Hydroelectric power, in particular, has played a significant role in this regard. Built between 1961 and 1965, the Akosombo Dam, also known as the Volta Dam, is a hydroelectric dam on the Volta River in southeastern Ghana in the Akosombo Gorge and part of the Volta River Authority. The construction of the dam flooded part of the Volta River basin and led to the subsequent creation of Lake Volta, one of the largest artificial lakes on Earth.[7] The primary purpose of the Akosombo Dam was to provide electricity for the aluminum industry, and it is considered Ghana's largest development project to date that has had significant

---

[7] Lake Volta covers nearly 3,275 square miles and is bigger than the land area of Rhode Island and Delaware combined.

human and environmental impacts (Gyau-Boakye 2001; Abrokwa-Ampadu 1984). In many ways, this dam project led to massive displacement and out-migration from the Volta Region and severely "disrupted the livelihoods of fishing communities and forced a significant proportion of the population to migrate to other parts of Ghana" (Yaro and Tsikata 2015).[8] All of this is to say that both colonial and postcolonial histories and policies, in combination with more recent economic development projects and natural resource management decisions, have impacted Ghana's internal migration patterns. As scholars of Ghana's contemporary political economy have noted (Songsore 2009; Awumbila, Teye, and Taro 2017), the forces of market liberalization—forces which ultimately determine prices, allocation of resources, capital investments, etc.—will only continue to further concentrate the power of urban centers in the southern region, largely because the greater Accra area continues to have the economies of scale and infrastructure needed to attract investors. It is these conditions that ultimately "reinforce existing socioeconomic differences and thereby sustain the rural to urban and north to south flows" (Awumbila, Teye, and Taro 2017, 986).

In addition to these conditions and struggles in the Volta Region, and certainly long before the scrap metal trade dominated economic activity in the Agbogbloshie and Old Fadama area, violent chieftaincy conflicts in the Northern Region during the 1990s and early 2000s forced many northerners to migrate to Accra in search of safety. It is estimated that roughly 10,000–15,000 people fled to Accra during what has become known as the "Konkomba–Nanumba conflict" or the "Dagbon chieftaincy dispute," which led to a civil war in 1994 (see Table 2.1). These tensions between the Kokomba and "majority tribes" (Dagomba, Nanumba, Gonja, and Mamprusi) have been the source of ethnic conflict research in the region (Assefa 2000; Bogner 2016; Talton 2016). These same "northern" ethnic rifts have even led to more recent interethnic violence and instability in Old Fadama (Stacey and Lund 2016). Northern Ghana is made up of over 30 ethnic groups that have traditionally been divided into chiefly and acephalous societies. The former have organized themselves around hereditary chieftaincy structures that have a hierarchy from lower-level chiefs to divisional chiefs, paramount chiefs, and even some who are superior to paramount chiefs and act like kings. As Rathbone (2000) has examined, the relationship between chiefs and political

---

[8] Other dam projects, like the Bui Dam along the Black Volta River, resulted in social displacements calling for "salvage archaeology" (see Gavua and Apoh 2011).

Table 2.1. Brief Timeline of Conflicts in Northern Ghana in the 1990s

Feb 2, 1994: Fighting in the north near the border with Togo broke out between Konkomba and Dagomba ethnic groups. The incident began with a dispute over prices in a market, but quickly accelerated to large-scale violence. The two groups have been at loggerheads for many years because the Konkomba, who are not Ghanaian natives, are denied chieftainship and land. Only 4 of 15 ethnic groups in the region have land ownership.

Feb 10, 1994: The government issued a state of emergency in the northern region (the districts of Yendi, Nanumba, Gushiegu/Karaga, Saboba/Chereponi, East Gonjo, and Zabzugu/Tatale and the town of Tamale). About 6,000 Konkomba fled to Togo as a result. The government also closed four of its border posts to prevent the conflict from spreading.

Mar 1994: The government fired on a crowd in Tamale killing 11 and wounding 18. Security forces fired on mainly Dagomba after they had attacked a group of rival Konkomba. It is difficult for the government to reach Konkomba fighters since they operate in small packets under bush cover. Members of the Dagomba, Gonjas, and Namubas (allies) turned in their arms in compliance with a government order to all warring factions. The seven districts affected by the fighting are the breadbasket of the region, and food prices have increased since the fighting broke out in February.

Mar 4, 1994: A grenade exploded in Accra in a Konkomba market injuring three. It is thought to be a spillover from the violence in the north between the Konkomba and Dagomba.

Apr 1994: An 11-member government delegation held separate talks with leaders of the warring factions in Accra. Both sides agreed to end the conflict and denounce violence as a means of ending their conflict. The 3-month-old conflict left over 1,000 (one report suggested 6,000) people dead and 150,000 displaced.

Jun 9, 1994: A peace pact was signed among all warring factions in the north. Two main groups of disputants were involved in the fighting (Konkomba vs. Dagomba, Nanumba, and Gonja), as were several smaller groups (Nawuri, Nchumri, Basari). No incidents were reported in the past several weeks, though the region remained tense.

Jul 8, 1994: Parliament agreed to extend the state of emergency imposed on the seven northern districts for a further month.

Oct 1994: Police seized arms bound for the north. The Tamale region is tense, and the peace agreement signed in April was regarded as a dead letter. Dagomba communities, backed by the Nanumbas and Gonjas, again began buying arms. Many Konkomba have been keeping out of sight following a series of lynchings.

Mar 1995: Renewed ethnic fighting in the north left at least 110 dead and 35 wounded. The Konkombas were largely blamed as instigators of the latest violence. The government had the situation under control by the end of the month. In Nanumba District, five arrests were made in connection to the violence. A total of 25 have been arrested since September 1994 in connection to the violence. Latest casualty figures put the number of dead at 2,000 since February 1994, and 400 villages and farms have been burnt to the ground.

Apr 1995: The government began proving funds for the rehabilitation of displaced persons from the ethnic conflict. An estimated 200,000 have been displaced. Most health, education, and water facilities were destroyed in the wake of the conflict and most personnel fled the area. Outbreaks of cerebro-spinal meningitis, polio, diphtheria, measles, tetanus, and whooping cough were reported. Agriculture in the area is nowhere near its preconflict levels.

Source: Minorities at Risk Project (2004).

leaders, especially relations forged between southern chiefs and Ghana's first president, Kwame Nkrumah, continues to be a source of cultural–political friction. Tensions between chieftaincy, democracy, and "development" were at the heart of these political relations; and in Ghana these complex relations endure in the postcolonial era.[9]

According to Rathbone (2000, 4), while diversity and variation characterize chieftaincies in Ghana, "[w]hat unites them all is their experience of an assault upon them which altered them in ways which went far beyond the shifts and adaptations which inhere in less traumatic historical change." An indigenous system of governance that persists in Ghana, these chieftaincies are made up of executive, legislative, and judicial powers that have endured alongside colonialism and imperialism. This chieftaincy system, with its own governing political economy, has also maintained significant control over land:

> Eighty percent of land in Ghana is held by the various traditional authorities in trust for the subjects of the stool/skin [symbol of chiefly authority] in accordance with customary law, and central government has ten percent for public development. . . . Chiefs are custodians of the resources within their various communities. In resource-endowed areas, as is the case with most of the stools in southern Ghana, chiefs exploit the resources for the general good of their communities. . . . Perhaps what is "new" is that chiefs are employing very innovative and seemingly modern means to achieve this goal. (Bob-Milliar 2016, 390–391)

There are four dominant chieftaincy-based ethnic groups in the Northern Region: the Dagombas, Nanumbas, Gonjas, and Mamprusis. The acephalous groups, such as the Konkombas, Nawuris, Basares, and Nchumurus, are segmentary societies that have not had hierarchical structures such as chiefs and chieftaincies. As Pellow (2011) has noted, the history of chieftaincies in Dagbon, or what is also called the Dagomba Kingdom, is no doubt complex. The kingdom was established in 1600 (Staniland 1975) and its Muslim

---

[9] As Pierre (2013, 60) adds,

The Chiefs Act of 1959 empowered the central government to have full control over recognizing or deposing chiefs (Boafo-Arthur 2007). Thus, from the beginning of full political independence in 1957 until its overthrow in 1966 [led by the National Liberation Council], Nkrumah's government "minimized the political and judicial role of traditional rulers, broke their financial backbone and made them passive appendages to the central government" (Brempong 2006:30).

majority, who speak an Akan language known as Dagbani, makes up about 8% of Ghana's total population. As Pellow (2011, 135) explains, the Dagomba have a social structure that is made up of individuals with both ascribed and achieved social status and power:

> The Dagomba have "big men", opinion leaders or patrons, who come by their position by combining ascribed and achieved statuses, although stressing the achieved—as engineers, medical doctors, lawyers, or parliamentarians. But "even certain 'officially' ascribed statuses (e.g., certain high-level inherited chieftaincies) are 'achievable' under the right conditions or with enough money or power" (Kirby 2003:180). So, for example, the chieftaincies (skins), which are only open to people of certain ascribed status, are ranked, and chiefly progression is channeled more and more narrowly until one rises to the skin (*nam*) of Yendi. . . . The higher chieftaincies are both ascribed and achieved. The kingdom's capital is Yendi, and it is here that the king, the Ya-Na, resides. The Dagomba are divided into two "gates" (clans), Abudu and Andani, each an offshoot of one of the two sons of Ya-Na Yakubu, who died in the late 19th century. (Staniland 1975; Tsikata and Seini 2004)

In most of the Northern Region, the traditional land tenure practice has not recognized individual ownership of land. Landownership, to a very large extent, has been vested in paramount chiefs and is held in trust or on behalf of the ethnic groups to which the chief belongs. This, therefore, has restricted landownership to chiefly groups. These groups argue that they were original settlers who allowed the acephalous groups migrating from other areas to settle on their land and farm there by permission. For this permission, the settlers pay tribute to the chiefs, although in many instances the tribute has become more and more symbolic. The acephalous groups have resented the monopoly of landownership in the hands of the chiefly peoples as well as the tribute that they are required to pay. Some of the acephalous people refer back to the sixteenth century to justify their claim that *they* were actually the indigenous people in the area and were invaded by the chiefly groups, who then took over the land and imposed their rule on them. The acephalous people insist on the creation of their own paramount chieftaincy that can hold land in trust for them.

Alongside chieftaincy and land entitlement politics are deep resentments based on perceptions of economic and political inequalities, social and cultural prejudices, and competition for limited resources. Additionally, the growth of multiparty politics in Ghana has made population growth a sensitive issue. For example, some acephalous groups have been increasing rapidly, leading to increased demand for representation in national and regional politics. This has generated broad concern among members of chiefdoms as this trend is considered a significant threat to "traditional" authority in the area, which is largely based on ethnicity and control of land. To further complicate issues, religion has also played a role in reinforcing conflict. While chiefdoms in Ghana's Northern Region are primarily Muslim, the leadership of acephalous groups is predominantly Christian; and many of these leaders have close ties with Western churches and missionaries, especially those who settled in and around Tamale,[10] a village (and now bustling city of roughly 250,000 inhabitants) that became the administrative headquarters of the British in 1907.[11]

Today, many young Dagomba migrate to Agbogbloshie and Old Fadama not to escape intertribal violence in their villages but to make wages to eat, to pay for school, and to help support their families. Like many regions of the Global South, Ghana is dominated by an "informal" economy that draws labor from many regions, both within Ghana and beyond (e.g., especially Burkina Faso and Nigeria). So even while Agbogbloshie can be depicted as a fiery space of toxic e-waste extraction, it is equally a node in a complex global network of informal labor migrations and socioeconomic organization. It is but one postcolonial zone in a "planet of slums" (Davis 2006),[12] an urbanizing planet where more than 1 billion people in the Global South live and work in informal settlements which now dominate an ever-expanding cityscape. Moreover, postcolonial urbanization across Africa in particular contains settler-colonial legacies that characterize African cities more broadly: "Even if we are now more than a half-century past the date of independence in

---

[10] In fact, a Catholic missionary in Tamale led to the creation of the Tamale Institute of Cross-Cultural Studies (TICCS) in 1971. TICCS "is a research and teaching facility and is a part of the Catholic Church of Northern Ghana. This is a place where you can expand your mind while focusing on your commitment to the Catholic faith. There are a number of different courses to choose from, with students and interns coming from all over the world to be a part of this amazing learning experience" (www.ticcs.com). Today, TICCS is led by a Ghanaian anthropologist who trained in Europe.

[11] For a more detailed history of political change in the Northern Region and Tamale throughout the twentieth century, see Staniland (1975) and MacGaffey (2006).

[12] Cities and the urban landscape have long been a central focus of numerous ethnographies exploring the interrelationship of poverty, slums, and marginality. See, for example, Hannerz (1980, 1996) and Burton (2005).

many African countries, contemporary cities on the continent still cope with colonial legacies in sociocultural and political-economic terms" (Myers 2011, 69).

This north–south migration directly or indirectly influences the lives of residents living in Old Fadama and working in and around the Agbogbloshie scrap metal market, which also includes women and children of various ages. Of the 50 men and women I surveyed and interviewed who work in Agbogbloshie, the frequencies and reasons for migrating to and from the north vary. What is for sure is that migration matters for most of these workers and that access to education, maintaining social networks, and participating in agriculture in the north are central to their lived migratory experience (see Table 2.2).

Those doing this migration usually ride one of the many buses that travel between the Northern Region and the coast. The ride from Accra to Tamale, takes roughly 12–14 hours each way, depending on the quality of the bus and, of course, luck. During fieldwork in the summer of 2017, I experienced a bus breakdown between Kumasi and Tamale that extended the trip to 17 hours. My field notes from that experience read "Today the bus broke down and it will only be a matter of time before this migration machine makes its way to Agbogbloshie as scrap metal!"

Others making the journey, use a motorcycle if they can afford to buy or rent one. Some workers travel back and forth form Accra to Bolgatanga, just near the border of Burkina Faso; but most are headed for Tamale, the so-called nongovernmental organization headquarters of Ghana and, not

Table 2.2. Reasons and Frequencies of Migration for Agbogbloshie Workers

"I go back every 8 months." (Abdurazak, 18-year-old male from Savelugu)

"I go every 2 weeks. I go back to farm." (Issah, 20-year-old male from Savelugu)

"I go back to school every 3 months." (Barkisu, 18-year-old female from Savelugu)

"I go and come back because my mother is in Old Fadama." (Amatu, 18-year-old female from Savelugu)

"I will go back to Savelugu in 2 months to be with my mother and father. I go with my 1-year-old girl." (Hawa, 19-year-old female from Savelugu)

"I go one or two times, if lucky." (Gafaro, 18-year-old male from Savelugu)

"I will go to Tamale in 2 months to be with my mother and father. I am 2 months pregnant." (Adija, 18-year-old mother of two, from Tamale)

Source: Worker survey conducted in Agbogbloshie, July–August 2017.

ironically, home of the country's University for Development Studies.[13] Many workers I interviewed and surveyed described how important school was to them, or at least how education figured centrally in their reasons to return to the north, even though this region has a deep history of formal educational neglect. For example, before World War II, mission-based educational programs in the Northern Region were not supported by the colonial government. While Cape Coast saw its first school open in 1876, it wasn't until 1951 that the Northern Region saw the development of its first secondary school, in Tamale (Pellow 2011, 136).[14]

Others I interviewed travel north to attend to family struggling with illness. For example, during a trip north in May 2018, one worker, Sadik, joined me to visit his father, who had recently experienced a stroke. He was, of course, saddened by the news that his father was suffering. Aware of this, several of the copper burners pitched in money to help him make the trip north. In fact, one worker who didn't want to pitch in, was shamed by the other workers. Also, several pregnant women I met told me that they prefer to travel north to be with family before they deliver. Memuna, after having her first child Abrahim, went north to live with her in-laws. A new mother with a quiet demeanor, Memuna bashfully explained "I go north now to be with his family. To learn how to raise Abrahim." I joined her on the trip north in 2016, after helping bring the baby to the nearby Korle Bu Teaching Hospital in western Accra to check the baby's growth and development. This was a special trip for Memuna, her husband Ibrahim, and their newborn. They were going north to unite Abrahim with his family. Ibrahim explained that "Memuna will stay north with my father and mother to learn how to care for the *pikin* [baby]. She will stay there." The baby did not make a sound during the entire 14-hour journey, and I caught glimpses of Ibrahim and Memuna looking at their baby, sharing a familiar smile as they experienced their new life as parents. Ethnography helps reveal the social intimacy of e-waste labor migrations, especially the social relations and practices of affection that keep people like Ibrahim and Memuna hopeful for a better life, and certainly a life

[13] Established in 1992, the university's stated mission "is to address and find solutions to the environmental problems and socio-economic deprivations that have characterized northern Ghana in particular and are also found in some rural areas throughout the rest of the country" (https://uds.edu.gh/).

[14] As Pellow (2011, 136) points out, "by 1925, there were only five government and two mission primary schools (Bening 1990). Clerical and skilled work was done by educated southerners, further underlining the North–South divide. The backwardness of the North was reinforced by officials who acted to prevent 'progressive ideas from the south' (Kimble 1963, 535) from creeping in (see also Grischow 2006)."

for Abrahim that is free of toxic smoke. As Ibrahim told me during our long bus ride, "Agbogbloshie is no good for him. The smoke is no good for him. The north is good for him."

But not all e-waste workers in Agbogbloshie engage in these north–south migrations. For example, one woman I met in 2017, who was from Bolgatanga, told me "I stay here because my family is here. I went to JHS [junior high school], above P5 [primary school]." Another worker explained, "I stay here. I live in Sabon Zongo [a Hausa settlement adjacent to Agbogbloshie] and completed JHS." Despite these cases, most of the workers I interviewed come from Savelugu, a village just north of Tamale and the home of my primary interlocutor, Ibrahim, who is a "chief" of a small group of copper wire burners in Agbogbloshie.

Life Between Agbogbloshie and Savelugu

I traveled from Accra north to the village of Savelugu for the first time in July 2016. It is a migration route that Ibrahim, my primary contact and interlocutor at Agbogbloshie, makes several times a year. Ibrahim Akarima, is a head copper wire burner who also happens to be a drummer for the paramount chief of Savelugu, the Yoo Naa. In Dagbon, it is said that a chief only has the power they have if they have drummers by their side.[15] Dagomba drummers speak the wisdom of the chief's lineage, and this therefore awards drummers a significant social status among the Dagomba. Dagomba drums are referred to as "talking drums," a term attributed to drums across much of West Africa. These drums (e.g., *lunga* and *gungon* drums) are not just musical tools but are used instead to speak literal words. In this way, language and drumming are deeply interconnected in Dagbon; and the Dagbani language, as a tonal language, is conducive to being played through drumming. Every chief in Dagbon has his own set of drummers. Traditionally, a chief would not walk or go anywhere without an entourage behind him, including

---

[15] As it was described to me, Bizung, son of Naa Nyagsi, was the first drummer in Dagbon. His mother died while he was young. As a motherless child, he struggled to survive. The only pleasure he found in life was to tap on a calabash drum outside the family compound. Over time, his drumming gained respect. Eventually, Naa Nyagsi offered Bizung the paramount chieftaincy. Bizung declined the offer and instead asked that he be allowed to play his music in peace. Naa Nyagsi agreed and appointed him as the court historian, and thus, the central role of the drummer in Dagbon began. All drummers today trace themselves back to Bizung, who they call "grandfather." For a fuller version of this story, see Locke (1990).

drummers. When a chief is on the move, drummers play to announce his presence and family lineage. So, Ibrahim, like many drummers before him, has a reciprocal relationship with chiefs in Dagbon. When drummers drum they convey an air of power to the chiefs, and drummers earn prestige as important figures in a chief's court. But for drummers like Ibrahim, there is a catch to his ascribed social status as *akarima* ("drummer"). In an interview with Ibrahim at the burn site in Agbogbloshie that he oversees, he shared that "The chief don't like me to travel. I am *akarima*, so he not like me to travel. All my brothers they be pray for me. They pray for me to do the drumming" (Figure 2.1).

Ibrahim, like other copper wire burners, lives a complex life on the move. Like other e-waste recyclers and urban copper miners, Ibrahim is on a life's path of work, fatherhood, and chieftaincy obligation; and this informs his north–south migrations. There are various ways in which he sustains his social ties in Savelugu, a village in northern Ghana where he is a key member of the local chiefdom. The Northern Region, which is Ghana's largest designated

Figure 2.1. Ibrahim drumming in Savelugu with his father. Photo by author.

region, is a massive area. It is bordered on the north by the Upper West Region and the Upper East Region, on the east by the eastern Ghana–Togo international border, on the south by the Black Volta River and the Volta Region, on the northwest by the Upper West region and Burkina Faso, and on the west by the western Ghana–Ivory Coast international border. Savelugu is the capital of Savelugu-Nanton, one of the 26 districts in the Northern Region. Like many villages in this region of the country, and given their proximity to the Sahel and Sahara deserts, the environment of Savelugu is semiarid; and the vegetation consists predominantly of grassland, especially hot savannahs marked by patches of drought-resistant trees such as baobabs or acacias.

Agrarian sector employment is a dominant source of economic activity in all regions of Ghana. According to Konadu and Campbell (2016a, 6),

> In all ten regions, most people farm, on either a subsistence or a commercial basis. More than half the total population is under age twenty-four, and though urbanization is on the rise, half of the population still lives in rural areas, 80 percent of these in villages . . . rural peoples are the backbone of the national economy . . . the agrarian sector, for instance, employs approximately 60 percent of the total Ghanaian workforce, which averages about 10 to 11 million. The agrarian and natural resources sector, with the exception of fisheries, produce cocoa (second in value only to gold as the largest export), rice, coffee, timber, industrial diamonds, manganese, bauxite, wood products, textiles, pineapples, cotton, plantains, and coco-yams for some indigenous markets, but largely for export.

But, even while farming remains an enduring source of employment across all regions of Ghana, it is increasingly the case that e-waste workers like Ibrahim migrate south to Agbogbloshie to make wages that can't be matched by farming in the Northern Region, especially since the late 1970s when Ghana witnessed significant declines in both rice production and investments in educational infrastructure (Antoine 1985; Pellow 2011).[16] Reflecting on her extensive ethnographic fieldwork in Ghana's Northern Region, anthropologist Deborah Pellow (2011, 136) explains,

---

[16] Educational inequities were at the center of national politics in the postcolonial era, especially within the context of north–south relations. As Pellow (2011, 136) notes, Ghana's first president, Kwame Nkrumah,

> beloved in the North, sought to rectify these inequalities by making education compulsory and, in the North, completely free. However, many of the new Dagomba elite I spoke with observed that they had gotten to school somewhat by accident. According to Aminu

In the 1970s, official economic policy was focused on agricultural improvement and, especially, rice growing, and indeed, the rice revolution and investments in cotton production did positively affect the region. Most Dagomba farmers who participated in rice farming did so as part of the state-sponsored rice boom. Because of the success of rice cultivation in the 1970s, some of those I interviewed in 2006 mentioned Ignatius Kutu Acheampong (a disgraced former head of state) as one of the two politicians (alongside Kwame Nkrumah) who had had the greatest positive impact in Dagbon and, indeed, the whole of northern Ghana (see Antoine 1985). Many Dagomba living in Dagbon today told me that what matters is that there was work for everyone. Sadly, by the late 1970s, it was clear that the country's agricultural program, and especially rice production, "had failed in its major objective of making the country self-sufficient in food." (Antoine 1985, 337)

Additionally, trade liberalization policies deployed across Africa in the 1980s as part of International Monetary Fund– and Work Bank–sponsored "structural adjustment" programs have put further stress on those struggling to sustain agricultural livelihoods. For example, since the early 1980s, sub-Saharan Africa has spent $272 billion to liberalize trade (Bond 2006). The downside is that such trade is based primarily on devalued agricultural commodities and therefore is "especially difficult to rely upon for growth, given that agricultural subsidies accruing to Northern farmers rose from the late 1980s to 2004 by 15 per cent, to $279 billion, mainly benefiting large agro-corporate producers" (Bond 2006, 159).

While commercial interests are a definite factor, subsistence farming for household consumption is a central focus of farming in Dagbon villages (Dickson 1968; Oppong 1973). For members of Ibrahim's family in Savelugu, a primary agricultural practice is groundnut farming, especially peanuts. During one visit to Savelugu in 2017, Ibrahim and I spent several hours hanging out with members of the village eating from a huge pile of peanuts, which is a staple of the local diet. Ibrahim asked me, "You like da *cema* [peanut]?" "Yes," I quickly replied, as it had become my favorite go-to snack during visits to the north. "We Dagomba eat *cema* all the time. We farm

Ahmadu, a civil engineer, everyone was surprised when his imam father sent him, his eldest son, to school at age ten; his maternal father's father felt sending him to school was risky "because anyone who went to school came out a Christian and was consuming alcohol, and that is very repulsive to our system."

cema. We eat many *cema*," Ibrahim added as he speedily shelled a handful of nuts and teased me about how slow I eat them. Despite enjoying eating groundnuts, Ibrahim and his fellow Dagomba struggle to rely on subsistence farming—mainly corn and peanuts—as a source of income (Figure 2.2).

Dagomba subsistence farming struggles are further compounded by climate change and broader burning truths. Longer droughts and more frequent flooding events only exacerbate contemporary agricultural livelihood challenges in the Northern Region and across Africa (World Bank 2019; Stocker et al. 2013; Tosca et al. 2015). As is the case in many tropical regions of the world, including Africa, farmers are burning biomass on their agricultural lands at an accelerated rate (Power et al. 2008; Archibald, Staver, and Levin 2012). This burning has been linked to various responses and changes due to land use, population density, and climate, especially in places like northern Ghana, which is among many vulnerable savannah ecosystems (Kloster et al. 2012; Moritz et al. 2012). Ghana's Northern Region has more recently been the focus of research connecting global warming, crop farming, and food security (Tahiru 2019). All of these processes expose the complex ways in which e-waste migrant labor in Agbogbloshie involves

Figure 2.2. Ibrahim's home and family in Savelugu. Photo by author.

broader environmental disturbances in Ghana that have led to significant agro-climatic migrations. But, in addition to these ecological drivers of many north–south labor migrations, there are kin and social networks and relations with deep chiefdom roots.

Amidst Chieftaincy and "Trans-Local" Life

Agbogbloshie and Old Fadama are urban spaces that are integrally linked to sociopolitical life in Ghana's Northern Region. This situation illustrates how urban–hinterland relations in Ghana are partially conditioned by "remote" chiefdom power, oversight, and control.[17] As Earle (1997) has noted, the power of chiefdoms comes from a variety of interlinked sources of social control and authority:

> In chiefdoms, control over production and exchange of subsistence and wealth creates the basis of political power. . . . Economic power is based on the ability to restrict access to key productive resources or consumptive goods. . . . The real significance of economic power may be that the material flows through the political economy can be channeled by chiefs to nurture or sustain the alternative power sources. (p. 7)

While these are certainly likely sources of the Yoo Naa's power, both in Savelugu and in Agbogbloshie and Old Fadama, Ibrahim has never explained the Yoo Naa's chiefly authority in these terms. During my first stroll through Old Fadama with Ibrahim in July 2015, he wanted me to see how chiefs matter in the social life of those living in Old Fadama. As we zigzagged through the muddy corridors of Old Fadama, we came to the seat of Chief Behi, who had recently been "installed"—the term used to describe chiefly selection—as a subchief to the Yoo Naa (see Figure 2.3).

Pointing at the chief installment poster, Ibrahim made sure I understood things correctly. "The Yaa Naa is for all, the Yoo Naa is number two. Yaa Naa is in Yendi. Yoo Naa in Savelugu. Chief Behi is now subchief here in Old Fadama." This calibrated celebration of "Chieftancy!!" is more than simply an illustration of the urbanization of chiefdom authority; it is a marker of

---

[17] For an in-depth analysis of the influence of Dagomba chiefdom power among Dagomba educated elite living in Accra, see Deborah Pellow (2008, 2016).

Figure 2.3. Chieftaincy public announcement in Old Fadama. Photo by author.

the recursion and extension of the political economy of "kingship" writ large (Graeber and Sahlins 2017). The Yoo Naa figures centrally in the power landscape of Agbogbloshie and Old Fadama for the very reason that his power is connected to the endurance of the kingdom and the practice of kingship itself. As Graeber and Sahlins recently put it,

> Kingship is a political economy of social subjugation rather than material coercion. Kingly power does not work on proprietary control of the subject people's means of existence so much as on the beneficial or awe-inspiring effects of royal largess, display, and prosperity. The objective of the political economy is the increase in the number and loyalty of subjects—as distinct from capitalist enterprise, which aims at the increase of capital wealth. (2017, 15)

Chiefdom loyalty is produced in a number of ways, but parades are common ways to display this loyalty. During a visit to the Yoo Naa's palace in Savelugu in 2016, I had the opportunity to sit with the Yoo Naa in his TV room, where he put on a DVD of a recently recorded parade. His assistant told me that he wanted me to see the DVD, which amounted to a 25-minute clip of the Yoo Naa being paraded through a village in Dagbon. I was told that the Yoo Naa wanted to show me an aspect of Dagomba culture, but what his assistant didn't tell me was that this is a primary way in which chiefs attempt to garner regional political power.

Agbogbloshie and Old Fadama are places where chiefdoms attempt to extend their socioeconomic power in a multiethnic and transnational scrap metal marketplace. This situation doesn't necessarily translate into Dagomba dominance because it is actually Nigerian businessmen who seem to hold the strongest economic position within Agbogbloshie as they make up the majority of scrap metal dealers.[18] But urban chiefs from Dagbon do oversee local social ties and relations between workers, even if they appear economically weak within Ghana's broader scrap metal trade economy. This became most clear to me when Ibrahim began to arrange a naming ceremony for his son Abrahim. In order to arrange the ceremony, Ibrahim and I met with several subchiefs in Old Fadama, a process that I later learned was a necessary step in "officially" approving Abrahim's naming ceremony.

Discussing e-waste labor in Agbogbloshie and social life in Old Fadama calls for an approach with a more complex political lens that recognizes on-the-ground power relations. Beyond the globally circulated pollution "victim slot" (Hughes 2013) storylines, we find that e-waste labor is tied to broader cultural politics and power relations that shape urban–hinterland relations. Surely, Agbogbloshie exists within Ghana's urban margins, but it is also a place that breaks down and confuses vocabularies of the urban and the rural. E-waste labor migrations inform the ebb and flow of socioeconomic life in Agbogbloshie, which provides a platform for rethinking urbanity in Ghana, especially the unexpected ways in which e-waste labor interconnects with and emerges alongside processes of "planetary urbanization" (Brenner 2018), a theory whereby cities are not the only nodes of urban extraction and waste.[19] Agbogbloshie is, in this sense, not simply an isolated scrap metal

---

[18] One possible explanation for this strong Nigerian presence in Agbogbloshie is the deeper history of social and economic relations between Ghana and Nigeria. Certainly, after Ghana became the first independent African state in 1957, many Nigerians moved to Ghana; and in the late 1970s many Ghanaians moved to Nigeria as economic migrants. But relations between Ghana and Nigeria became contentious when in the early 1980s Ghana began to enforce stricter immigration laws, resulting in mass deportations. Nigeria retaliated in 1983 by deporting over 1 million Ghanaians, at a time when the country was struggling with severe drought and economic depression.

[19] As Brenner (2018) explains,

This situation of planetary urbanization means, paradoxically, that even spaces that lie well beyond the traditional city cores and suburban peripheries—from territories of

market but instead a site of interconnection conditioned by a planetary urbanization process that can make certain urban places and spaces vulnerable to e-wastelanding and e-waste burning.

Ibrahim and his tight social network of copper wire burners navigate this "worldwide urban fabric," but their urban margin experiences and stories are closely threaded into their rural livelihoods and chieftaincy bloodlines. Nearly everyone working in Agbogbloshie maintains some social tie to kin in Ghana's Northern Region. Therefore, laboring bodies and migrations in and out of Agbogbloshie are integral to the globally circulating story of Agbogbloshie as one of the "most polluted place on Earth," as Pure Earth would have it (Pure Earth 2015). Surely, it can be discussed in terms of an object of e-waste industrial pollution and risk, but Agbogbloshie is also a site of fluidity, capital, and hope. It can also be thought of as a magnetic market of copper extraction and capital accumulation that actively links the rural and the urban in new and complex ways. Turning to the experience of e-waste migratory laborers themselves is, of course, integral to knowing anything about these labor migrations and rural–urban relations. These narratives are critical to also understanding what is actually involved in these migrations and relations and even what is expected of e-waste migrations as viable and sustainable sources of wage labor.

As anthropologist Dagna Rams has recently noted, the "trans-local lifestyles" of these migrant laborers are, in many ways, not only processes of circulating between urban spaces like Agbogbloshie and Old Fadama and villages in the northern regions but also practices that involve the "joining of two spaces—urban and rural, slum and village—[that] makes the urban centres into zones of economic mobility for individuals and development banks for the North" (Rams 2018, 1). These shifting rural–urban relations generate significant urban governance struggles, a situation found in other rapidly growing West African cities like Lagos, Nigeria, where toxic e-waste extraction is also widespread (Lawal 2019). As in postcolonial Accra, urban governance politics in Lagos is also shaped by similar tensions between rural ("traditional") communities and city ("modern") inhabitants who each seek

agro-industrial production, zones of industrialized resource extraction and energy generation, "drosscapes" and waste dumps, transoceanic shipping lanes, transcontinental highway and railway networks, and worldwide communications infrastructures to alpine and coastal tourist enclaves, "nature" parks and erstwhile "wilderness" spaces such as the world's oceans, deserts, jungles, mountain ranges, tundra and atmosphere—have become integral parts of a worldwide urban fabric. (p. 189)

to gain something from "new kinds of social and economic aspirations fostered by capitalist urbanization" (Gandy 2006, 380). So city–hinterland relations and chiefdom political culture and power figure squarely in the social relations making up the Agbogbloshie workforce, but these regional relations are nothing really new in Ghana. It is also worth remembering that before the renaming of the country in 1957, Ghana was called the Gold Coast. This coast, along with coastal communities and markets, and the hinterland savannah of the Northern Region have long been interlinked with systems of transatlantic slaving: "peoples from northern Ghana, such as those in Dagomba and Salaga, were a major source of captive peoples for the more powerful forest-based polities, and quite a few of them would join those procured along the coast and its hinterlands for a one-way voyage across the Atlantic" (Konadu and Campbell 2016b, 82). In this way, it is with informed caution that one might look at contemporary e-waste salvaging labor and global copper trades as deeper, more complex practices emerging from a Ghana with a much more brutal slaving history that involved violent extractions of value and unquestionable politics of survival.

Beyond Mere Survivalism

Hanging out in Agbogbloshie and spending entire workdays with e-waste burners opened my eyes to what other ethnographies of waste labor in the Global South have revealed. For example, my experience in Agbogbloshie, while different in context, region, and population, almost directly mirrors findings and experiences emerging from discard ethnography in Brazil. Working among *catadores* (self-employed workers) in Rio de Janeiro's largest dump, Kathleen Millar (2018) reflects on being perplexed by a general lack of questions of *why* workers return to do the work they do: "I have never been asked why catadores keep going back to the dump. Over time, I realized that this question is never asked because the answer is assumed. That is, the seemingly self-evident explanation for why catadores collect on the dump is that they do so out of necessity, as a means of survival. . . . The story ends before it begins" (p. 3).[20] This is shockingly exactly my experience in

---

[20] See Rosen (2020) for a detailed analysis of "sacrificial labor" in Ghana, especially as it relates to workers engaged in illegal small-scale gold mining, also known in Ghana as *galamsey*.

Agbogbloshie. Ghana, it turns out, was where the very notion of "informal" labor emerged (Hart 1978). As a space of informal labor options, one could explain Agbogbloshie in terms of informality, but the problem is that such an explanation can smack of essentialism. Under an informal labor lens, Agbogbloshie "appears as an end zone in a double sense: the burial grounds for unwanted things, the end of the line for urban poor" (Millar 2018, 4).

Struggle in Agbogbloshie mirrors many of the same struggles found in narratives of the Global South emphasizing slum life, resource scarcity, and situations of extreme precarity (Stacey 2019).[21] The working poor experience in Agbogbloshie is one that is beyond what Mike Davis (2006) calls "informal survivalism" because these workers do struggle to live the good life. In my interviews with them, they all told me they do this to earn "chop" money—or "food" money—but that doesn't tell you everything about why they are in Agbogbloshie or how their "northern" social networks extend to the urban margins of Old Fadama. There is more connecting labor and life than mere economic survivalism. Many workers return to Agbogbloshie because they have key members of their family who remain in Old Fadama. Ibrahim's "second"[22] mother Samatu, his sister Abuu, and his older brother Abrov are in Old Fadama. His father and his responsibility to the chiefdom and the Yoo Naa are in Savelugu, but these other core members of his family are in Old Fadama. Also in Agbogbloshie and Old Fadama are his friends, his "brothers," who he is eager to be close to. In fact, during my multiple visits to Savelugu it was apparent that Ibrahim was eager to return to Agbogbloshie, as if he missed the excitement of Accra. In Savelugu, where most of the workers I have interviewed come from, there is little opportunity to find work that pays more than 5 cedi a day. As noted in Table 2.1, the ethnic conflicts in northern Ghana in the 1990s informed the southward migration to Accra. These migrations led to the growth of the Old Fadama settlement, but in turn these migrations also increased the population of workers at the Agbogbloshie market, and therefore increased competition among workers. It has been noted by several studies of domestic migration in Ghana that finding work is not generally the sole driver of migration for these Ghanaians (Yaro et al 2011; Awumbila, Teye, and Taro 2017). Social networks play a

---

[21] As Azmanova (2020) notes, it is also important to recognize how "precarity capitalism" is shaping the experience of many communities, not just the poor and marginalized.

[22] Ibrahim refers to Samatu as his "second" mother as his father is a subchief with two wives. Samatu is his biological mother, but Ibrahim refers to her as his second mother because his father's first wife is the "first" mother of the family.

key role in these decisions, as illustrated in the case of Ibrahim who continually returns to Old Fadama to spend time with his family (Figure 2.4). His e-waste labor directly supports their livelihoods, a practice of distribution experienced by other copper wire burners like him.

Domestic migration studies in Ghana, for example, generally find that upon deciding to travel or relocate to the bustling capital city, "migrants then fashion out various strategies to enhance their economic survival in the large urban space of Accra. Many of these strategies are based on a complex intersection of social networks, which provide some source of social capital to counterbalance the disadvantages that migrants may encounter in Accra . . . migrants do not move blindly to cities in search of jobs, but are often assisted and supported in making decisions, travelling, settling, finding jobs, solving problems and optimizing livelihoods" (Awumbila, Teye, and Taro 2017, 986–987). In fact, one study found that of 1,500 households close to 60% of migrants leaving those households return and report having had a contact person in their destination before deciding to migrate (Awumbila, Teye, and Taro 2017).

Figure 2.4. Ibrahim with his sister Abuu (left) and mother Samatu (right). Photo by author.

If social networks help facilitate migrations to and from Agbogbloshie, it is equally important to recognize the ways in which my own relationship with Ibrahim and his family informed these south–north migrations. For example, during each fieldwork visit to Ghana, it was expected that Ibrahim and I would make a trip to Savelugu to greet his family there and meet with the Yoo Naa. A noticeable pattern of these short visits—usually 2–4 days—was Ibrahim's excitement to return to Agbogbloshie as soon as possible. The lure of the possible earnings was a definite "pull" factor that I witnessed each time we went north, reinforcing the economic, yet also temporal, dimensions of these e-waste labor migrations.[23]

The point is that Agbogbloshie's optimistic Dagomba migrants, like Ibrahim and other copper wire burners from Savelugu, live complex lives. Reducing their migration story to metal extraction and urban e-waste economies misses actually existing social realities that inform these migrations in and out of Agbogbloshie. As will be explored in the next chapter, the experience of domestic e-waste labor migration in Ghana is also an experience of everyday confrontations with urban displacement, land management violence, settlement demolition, and aggressive marginality. Dagomba making a life in Agbogbloshie and Old Fadama witness enduring threats and risks of displacement and marginalization. This experience, in turn, keeps them caught up in a constant situation of uncertainty and elusive survival in the face of aggressive urban governance. The next chapter interrogates the ongoing demolition logics and politics of erasure unfolding in Agbogbloshie and Old Fadama, showing how these local confrontations and governance situations in Accra's urban margins have a particular violent and ruinous pyropolitical dimension.

---

[23] For an ethnographic analysis of the complexities and consequences of "return" in labor migration studies, see Oxfeld and Long (2004).

3

Erasure, Demolition, and Violent Obsolescence in the Urban Margins

Look here. They come and burn my spot down. They no care. Nothing for me here. Why?

—Alhassan, Agbogbloshie worker

We'll eject Agbogbloshie squatters despite protest.

—Accra Metropolitan Assembly

At the heart of Agbogbloshie's complex narrative are matters of materiality, destruction, and affect.[1] While the Agbogbloshie scrap market is a place in constant flux and transformation—from goods to people to discard—it is also a site of aggressive reordering, infrastructural decomposition, demolition, eviction, and violence. For sure, remains of electronic destruction and obsolescence are everywhere in this e-wasteland, but a particular form of lived material destruction and disruption is also alive. In this chapter, I highlight how e-waste workers in Agbogbloshie have faced repeated rounds of eviction and forced displacement that have left them to navigate a dismantled and charred rubble life. Worker shelters have been routinely destroyed by urban authorities doing urban governance. Just as roads around and paths through Agbogbloshie are active forms of infrastructure on Accra's urban landscape, worker shelters, including the lasting rubble of their erasure, are symbolic "barometers of the lived experience of global political economy" (Harvey 2018, 86). Or, as Povinelli (2019) might put it, it is capitalism as usual: "Capitalism depends on creating by destroying and then

---

[1] Materiality, while central to science and technology studies and political ecology scholarship (see Latour 1993; Bennet 2010), is actually witnessing perhaps its greatest theoretical overhaul with the help of a growing number of discard studies scholars (Isenhour and Reno 2019; Liboiron 2016).

Burning Matters. Peter C. Little, Oxford University Press. © Oxford University Press 2022.
DOI: 10.1093/oso/9780190934545.003.0004

erasing the connections between the material wastes it leaves behind and the glimmering oasis of privilege this waste affords." In this situated context, however, the lived experience of the global political economy and ecology of electronic discard is further compounded by experiences of fiery erasure and infrastructural displacement. To make sense of embodied responses to extreme marginalization and "perpetual displacement" (Saethre 2020) in Agbogbloshie and the Old Fadama slum, I can't help but think about the force of "obsolescence" itself, a term that enflames the logics of "planned obsolescence" in our high-tech age.[2] Therefore, my intervention here is one that is informed by a situated urban political ecology of fiery erasure that develops what I shall call *violent obsolescence*, that is, the direct displacement-based experience and suffering of laborers engaged in a global copper supply chain fueled by an electronics industry that actively values material obsolescence.

In addition to toxic fires and exposures, these workers struggle to negotiate state-based forms of violent erasure fueled by demolition logics that paradoxically redirect and reorient the e-pyropolitics of environmental health in Agbogbloshie. In this way, copper wire burners experience marginalization in many ways, but repeated dwelling demolitions are experienced as an explicit form of "infrastructural violence" (Rodgers and O'Neil 2012). While "Infrastructure can be a key means through which social improvement and progress is distributed throughout society. A key conceptual challenge, then, is to understand when it is that infrastructure becomes violent, for whom, under what conditions, and why" (Rodgers and O'Neil 2012, 402). Attuned to the broader ethnography of toxic colonialism and postcolonial infrastructural violence more specifically, I argue here that worker experiences of displacement, eviction, and suffering[3] in Agbogbloshie and the Old Fadama slum provide a critical counterpoint to the intersecting and powerful logics and discourses of urban management, demolition, decontamination, and infrastructural optimism. I also attend here to how "environmental suffering" and "living toxicity" (Auyero and Swistun 2009), as experienced by copper wire burners living amidst these ground-level forces, sharpen our understanding and critique not only of Agbogbloshie's purported e-waste problem but also of how dispossession itself is experienced by Dagomba scrap metal

---

[2] For an excellent overview of the emergence and strength of the concept and practice of "planned obsolescence," see Slade (2006). Also, see Campbell (1992) for a discussion of how obsolescence is more generally fueled by a consumer culture of "neophiliacs."

[3] Following Auyero and Swistun (2009, 158), I am "deeply aware of the moral and political dilemmas revolving around attempts to represent the suffering and the domination of others," but I take the risk because I believe silencing this human experience does more harm than good.

workers in a postcolonial Ghana with an Environmental Protection Agency (EPA) that is cognizant of the poverty–pollution relationship and appears committed to turning environmental problems into economic opportunities. As one EPA officer I interviewed put it, "It is the goal of the government to not forcefully remove workers in the e-waste value chain." But ethnography is about digging deeper. Listening to the experience of those in this "e-waste value chain" reveals an experience of removal and injury that should not be silenced.

Witnessing Erasure and Demolition Pains

I returned for my third trip to Agbogbloshie in early July of 2017. On my first day back to the scrapyard, I met with Ibrahim and Jacob, another copper wire burner from the village of Diari just north of Savelugu, who came to work in Agbogbloshie in 2012. It was late in the afternoon, and we met at a spot in the scrapyard the workers call "Gaza," which functions as a shelter for workers who oversee the dismantling of junked trucks and focus their labor on the extraction of scrap iron and steel in particular. Before I arrived at Gaza to meet Ibrahim and Jacob, I passed by the old worker shelter the burners called "Bombay," a shelter and workplace made up of two structures near a live-stock holding station and water tank that had only a few weeks earlier been demolished by the primary urban waste management and infrastructure authority, the Accra Metropolitan Authority (AMA) (see Figures 3.1 and 3.2). This was the worker shelter where I spent most of my time doing fieldwork. It was where I hung out during the summers of 2015 and 2016, and it was there that I also met baby Abrahim, Ibrahim and Memuna's first child. The concrete blocks that made up the floor of the old shelter were still visible, but all else was gone. All that remained were indentations, shapes in the mud that had hardened. The scene was one of remnants and reminders of *place-making*, a common thread in ethnographic studies of marginalized migrants (Komarova and Svašek 2018; Low and Lawrence-Zúñiga 2003; Cresswell 1996, 2010). It was equally a scene of social memory, a phantom home for a group of Dagombas making a life in this scrap metal landscape.

The destruction and leveling of this e-waste worker shelter reminded me that Agbogbloshie is and continues to be a place and space of infrastructural transformation and social rearrangement with demolition machinery, fire, smoke, and power at its core. These ongoing changes are connected to

Figure 3.1. The Bombay shelter before the AMA demolition. Photo by author.

Figure 3.2. The Bombay shelter after the AMA demolition. Photo by author.

the lived struggle and experiences of displacement and marginality. Take the example of Jacob Agana, a former resident of Bombay. The tallest of his fellow workers, Jacob stands out in size and character. When I arrived in Agbogbloshie several months after Bombay had been leveled in the summer of 2017, Jacob sat with me at their new work area and explained what had happened to Bombay.

"So, what happened to Bombay? Was it those cows again?" I asked, as I had remembered that during my visit in 2016, the workers told me that the bulls roaming around the scrapyard occasionally would plunder their site. In fact, on an early morning in July 2016, I arrived at Bombay to find Ibrahim, Jacob, and a few others arguing over how best to rebuild their shelter. I was told that a bull knocked down the wood frame to their structure, which was not uncommon, especially since Bombay was located right next to a group of penned bulls. But this time, it was not the roaming bulls of Agbogbloshie that were the agents of destruction but instead the AMA.

"The AMA came there," Jacob quickly replied. "They say go. You be go. They say take everything away. The AMA has machines. The AMA use machines to demo. It was early in the morning. Like 4am. Ibrahim woke up and told us boys we had to go. All of us go. Abrahim, Rashida, Ibrahim, Memuna, Sadik, Issaka, and me. We go because da AMA. The AMA no good. They kick us out. Now we here. New spot is not good. *Kom* ["water"] be bad. New spot no good with da *saa* ["rain"]. *Saa* not good here. Bombay be better. Why the AMA do dat? No. Not good for us here."

Ibrahim, who also sat with us, interjected.

"This be the AMA. All the time. They do this to us here," Ibrahim added.

"Why did they come and destroy Bombay," I asked.

"They no like us staying here. They say you only work here. No stay here. The AMA say that," recounted Ibrahim.

Bombay was located along a primary footpath that everyone, including those who simply use the scrapyard as a shortcut to navigate this section of western Accra. I was curious to know whether they were relocated because they burned right on a primary footpath that many, even hundreds, of people used each day, but neither Ibrahim nor Jacob seemed to know why.

"Did they move you because of the smoke on the path coming from burning the copper wires?" I asked.

"I don't know. Everyone gets da smoke here. The AMA move us here. This I know," Ibrahim exclaimed as he loudly blew his nose and continued to give orders to his burner boys to go collect copper to burn. He reassured me, "The AMA always make problem for us."

The Bombay destruction in 2017 was not my first experience witnessing violent erasure, displacement, and demolition of worker shelters in Agbogbloshie. Other tactics of destruction, erasure, and displacement involve the explicit use of fire. As noted in the book's introduction, Agbogbloshie is made up of a variety of labor sectors. One group of workers, *bola* ("garbage") workers, roam the scrapyard looking to collect aluminum cans and plastic bottles. This is the work focus of Alhassan, who I met in 2016. Alhassan was 20 years old and the father of a 7-month-old girl when I met him. Ibrahim introduced us one morning, and he was excited to show me his recent aluminum and plastic collection (see Figure 3.3). He opened a bag of aluminum cans he had collected and proceeded to show me the shelter where he sleeps.

"Look over here," Alhassan said, grabbing my forearm to show me his shelter. He laid down in his bed so that I could see that it was comfortable.

"You see. Good cushion," he shouted from inside the lopsided shelter.

Figure 3.3. Alhassan's shelter before the AMA ignition. Photo by author.

"Nice. You look comfortable," I said, quickly realizing "comfortable" was likely an incongruous word in this situation.

"I sleep right here. My work is here," he added with excitement.

"Wow! You sleep and work here then?" I asked, while finding it hard to believe and experience for myself.

"I work *bola*. I grab aluminum, plastic, whatever I can sell here," Alhassan exclaimed.

He proceeded to organize his collection of cans and bottles, as I followed Ibrahim to collect a bundle of copper wires that had just arrived at his burn site.

On a cloudy morning about 10 days after visiting Alhassan's shelter, I witnessed the largest plume of dark smoke my eyes had ever seen (see Figure 3.4). As I walked to meet Ibrahim and the other copper wire burners, I could not take my eyes off this massive burning scene in front of me.

When I got to Bombay, Alhassan was sitting with Ibrahim and Jacob. When they saw me, Alhassan jumped up, pointing in the direction of the fire.

"You see the smoke, the fire?" he inquired.

"Yes. That is crazy. What is on fire?" I anxiously asked.

Figure 3.4. Alhassan's shelter after the AMA ignition. Photo by author.

"They burned down my shelter," Alhassan replied.

"Who burned it down?" I asked.

"The authorities. The AMA. They no love me," he angrily responded.

These experiences of eviction and forced displacement in Agbogbloshie and Old Fadama, which date back to colonial and precolonial periods, mimic what other anthropologies of displacement and erasure have found, namely that there is a lived temporal and placed-based dimension to the experience of erasure, displacement, and eviction. For example, Harms (2013, 344) points out that "Eviction thrusts many residents into an alternative time-world of enforced waiting, marked by an oppressive sense of being suspended in time."[4] The narratives of urban eviction and displacement explored here reveal how "the temporality of eviction produces different effects on different people" (Harms 2013, 347). Workers in Agbogbloshie and those dwelling in Old Fadama experience erasure and displacement in different ways, but what links their experiences is their shared exposure to repeated "infrastructural violence" (Rodgers and O'Neil 2012). Worker experiences reaffirm the "shadow" realities of neoliberal Africa (Ferguson 2005) in general and more specifically the violent erasure of situated scrapyard infrastructure in Ghana. These images and narratives communicate the usual cascade of problems confronted in Africa and other regions of the Global South where so-called development optimism is overshadowed by infrastructural neglect, social suffering (Ferguson 1999; Gandy 2006), and more chronic postcolonial "politics of capacity" (Tousignant 2018). Ethnographically, these images and narratives help expose the violent wastelanding and fiery displacement and dispossession that occurs in this West African "technocapital sacrifice zone" (Little 2017). More to the point, "Violent neoliberal political economies congeal in the city's wastescape and are made manifest in crisis moments" (Fredericks 2018, 13).

In Accra, the leveling and torching of worker shelters is an explicit act of destruction of "vital infrastructures of labor" (Fredericks 2018) by key urban authorities that, ironically, are mandated by the Ghanaian Parliament to oversee and manage waste, the "environment," and urban land use. In fact, the burners built Bombay only after they had been forcefully evicted from their dwellings in Old Fadama by the AMA, which claimed they were

---

[4] Alongside these embodied experiences of temporal suspension, eviction experience is also more generally conditioned by chronic experiences of poverty and property management politics (see Desmond 2016).

occupying a risky flood zone. Whether or not the AMA's decision to remove people from the flood zones was justified, the AMA was surely ready to use force against people who squatted where they were told they couldn't.

Planned Demolitions and Flood Control Conflict

Even though Agbogbloshie is an old neighborhood in western Accra that has become internationally recognized as an e-waste dumpsite, it cannot be understood without direct reference to a large informal settlement, Old Fadama, which lies just a few hundred meters to the southeast of the copper burner site. While Agbogbloshie and Old Fadama are geographically separated by the Abose-Okai Road, where a large majority of urban food is sold, these places are understood by all the workers I interviewed to be not only interconnected but extensions of the same community. Their interconnection, in fact, helps support the claim that Agbogbloshie and Old Fadama make up not only Ghana's largest urban slum but one of the largest in West Africa. These entangled urban settlements have attracted large numbers of migrants since 1981, but the majority have certainly been migrants from the Northern Region, like members of Ibrahim's work cohort. These informal urban settlements are also, not surprisingly, neighborhoods in Accra with the cheapest rents (Grant 2006) and risk being subjected to "parasitical" rent extraction (Gandy 2006).[5] But Agbogbloshie and Old Fadama are also informal and stigmatized zones of Accra targeted as harmful sources of pollution, illegal squatting, and violations of urban flood control policies. All of these factors inform the unraveling of demolition projects and logics in Agbogbloshie and Old Fadama and expose an ongoing urban political ecology of erasure writ large, and lagoon flooding management politics in particular.

As geographer Grace Abena Akese and I explore elsewhere (Little and Akese 2019), flood control and management decisions along the Korle Lagoon, the primary waterway that surrounds Old Fadama and the Agbogbloshie scrap metal market, plays a central role in eviction and demolition politics in the area. The Korle Lagoon is a place and space of intensive land use, toxic waste disposal, social life, and urban ecological restoration.

---

[5] This urban rent extraction is, of course, informed by a broader globalization of rentier capitalism that feeds on, among other things, urban land grabs and so-called ground control rents (Christophers 2020).

The banks of the lagoon house an ever-growing informal settlement, used as a solid waste landfill site and zone targeted by international nongovernmental organizations (NGOs) as a lagoon inundated with toxic e-waste runoff from the open burning of e-waste found at the Agbogbloshie scrap market. Currently, there are attempts to rehabilitate the lagoon through the Korle Lagoon Ecological Restoration Project, an ecological science and restoration project focused on the Korle Lagoon and its river system in the metropolitan area of Accra. This project has moved forward with efforts to redredge the lagoon and the in-flowing river system to restore drainage, sewerage, sanitation, and urban planning goals that fit within the sustainability scope of the project. In this way, the lagoon restoration project showcases the neoliberal complexities of ecological restoration in a multiuse marine environment in Ghana.[6] But it also involves direct linkages to worker evictions and dwelling demolitions.

Located to the southwest of the central business district of Accra, the Korle Lagoon is the major runoff water receptacle for the city of Accra. All the major drainage channels in the city are connected to the lagoon, which ultimately drains into the Atlantic Ocean at the Gulf of Guinea. Together with parts of its largest tributary, the Odaw River, the lagoon covers a total surface area of 0.6 km^2 and receives water from about a 400-km^2 catchment area (Boadi and Kuitunen 2002; Karkari, Asante, and Biney 2006). With increasing urbanization and a population explosion in Accra, about 60% of the Accra metropolis is within the catchment area of the Korle-Odaw basin (Karkari, Asante, and Biney 2006) making the lagoon significant in the natural hydrology system of this city of over 4 million people. Indeed, the challenges of urban flood management in Accra are intricately linked to the functioning of the Korle Lagoon (Amoako 2016; Amoako and Inkoom 2018). In the decades following independence and especially since the 1990s, the Korle-Odaw basin has become known as a heavily polluted landscape, an environmental disaster in need of mitigation. Historical evidence suggests that up until the 1960s the lagoon supported commercial fisheries with other attendant socioeconomic activities for the indigenous communities around its vicinity (Boadi and Kuitunen 2002). A thriving fin and shellfish trade occurred along the lagoon and served as a significant source of income for the nearby Ga villages and subsequent "native towns" established under the

---

[6] As Little and Akese (2019) describe elsewhere, this marine contamination issue exposes the "blue political ecologies of e-waste" in Ghana.

colonial administration. The lagoon is also remembered as a hub of canoe transport for traders. Importantly, it was also a reserve for aquatic biota including serving as habitat for major populations of wading birds (Ntiamoa-Baidu 1991).

A long and complex history of intensive land use and associated pollution accounts for the current state of the Korle Lagoon (Little and Akese 2019). As the major receptacle of runoff in the city, some of the earliest gutters and sewer pipes first laid to drain the city all connect to the lagoon for final discharge into the sea. In the absence of regular upgrades, however, and with increasing volumes of discharge coupled with the open nature of the drains, leaks and overflows from sewers are a primary cause of the lagoon's contamination (Onuoha 2016). Until very recently, when the "Lavender Hill"[7] fecal waste treatment site was operational, untreated fecal waste was directly emptied into the ocean by the city authority, the AMA), which backwashed into the lagoon, compounding its contamination. Additionally, the banks of the basin are lined with industrial activities including a paint factory, breweries, textile factories, garages, and vehicle repair workshops, which all discharge their effluent into the lagoon (Karkari, Asante, and Biney 2006). Effluent (including mortuary wastes) from the nearby Korle Bu Teaching Hospital is also discharged into the lagoon. Uncontrolled discharges, open defecation, and dumping of waste directly into the lagoon also occur from the ever-growing high-density low-income settlements within its vicinity.

Furthermore, in the absence of garbage collection services, household and industrial waste from the Old Fadama informal settlement and the adjacent Agbogbloshie scrap yard also spill into the lagoon. Of special note is the practice of land reclamation along the lagoon. A large portion of the Agbogbloshie scrap yard and Old Fadama sits on land that is largely part of the lagoon. Over time, and as the settlement continued to grow, residents resorted to filling in the lagoon with sawdust to reclaim land and build further at the edge of the lagoon (Farouk and Owusu 2012). This reclaimed land caves in frequently, releasing the sawdust into the lagoon. Contrary to the popular narrative attributing the "death" of the lagoon to the marginalized Old Fadama informal settlement and the economies it had established along the lagoon (prominently featuring e-waste processing), pollutants from Old Fadama and the Agbogbloshie scrap yard at the upper reaches of the lagoon

---

[7] The area is sarcastically known as "Lavender Hill" because of the emanating stench from the fecal discharge into the sea. Shortly before the Lavender Hill site was abandoned in 2016, a nearby four-star hotel was open for business.

are only a recent addition to the growing list of pollution sources in this area of Accra.[8] The sources of the lagoon's pollution are varied and importantly predate the adjacent e-waste industry along its banks, which only came into prominence around 2008–2009 (Little and Akese 2019).

Still, regardless of the pollution source studies available, the Korle Lagoon has been anchored to a "pollution nightmare" narrative. Here is one example from a journalist writing about the lagoon just around the same time the Agbogbloshie e-waste conflict caught the attention of Greenpeace International:

Theo Anderson, a Ghanaian ecologist, moved gingerly up to the edge of the pond, and remarked in a friendly way: "If you fall in there, you'll be dead in minutes." He wasn't joking. The Korle Lagoon in Ghana's capital of Accra is one of the most polluted places on earth. It is a natural depression that serves as a cesspool for most of the city's industrial and human waste—it is an environmental nightmare. Owing to the pollution, no living thing, animal or plant, has been able to grow in it for years. Even boaters steer clear of its thick, black nauseating syrup. Its stench wafts back to envelop the adjoining shantytown that is home to hundreds of families who, because they have no sanitation facilities, have turned the shores of the lagoon into a giant latrine. (Bourgoing 2002)

The Korle Lagoon is not only a multiuse marine environment and sink for toxic water, vegetation, and noxious scrap metal economies but also a place and space of social and environmental memory, a site of green memory-making (Little and Akese 2019). In fact, when the first Ghanaian environmental journalist, Mike Anane, exposed the Agbogbloshie e-waste problem, he was very open about the linkage between his environmental advocacy and his own eco-biography and environmental memory of the lagoon's once lush ecosystem. As he recently put it, "What was once a green and fruitful landscape is now a graveyard of plastics and skeletons of abandoned appliances" (Ottaviani 2018). In one interview with a middle-aged resident of Sabon Zongo, a Hausa-speaking community adjacent to Agbogbloshie with a deep

---

[8] This is not in any way to understate the damaging effects of e-waste processing on the lagoon. Obviously, there are numerous and highly toxic chemicals and materials (lead, cadmium, polychlorinated biphenyls, polybrominated diphenyl ethers) contained in e-waste, which compounds the problem. Instead, as Little and Akese (2019) point out, it is important to draw attention to this ongoing zone of socioecological contamination at play when looking at the broader urban economies facilitated by the lagoon.

social presence in this area of Accra, the lagoon is remembered as a childhood play area and place of adventure: "It was so green and we used to chase and trap rats. We were kids. It was a big place with so many trees. Nothing like it is now. So much has changed. I think dat is how it is bad. No more green. Now all there is is waste and smoke. So much has changed there."

The trashing of the lagoon and the accumulation of toxic heavy metal runoff from e-waste processing and disposal of organic material is also an underlying stimulus for local ecological restoration projects and marine toxicological studies of the lagoon (Hosoda et al. 2014; Huang et al 2014). The impact on marine life was a common theme of interviews conducted with residents in the surrounding area, many of whom recreate along the Korle Lagoon: "Many people run along this part of the Ring Road [a major road that arcs around Accra]. We run to the ocean that is at the end of the lagoon. That is where the fishermen are, and you see their boats in the black water. It is so so polluted. You know the smell. [Laugh]. When I was a boy, we would fish here. Now look at it. And you see the boats. They still fish in it. Everyone eats pollution here, even if the lagoon is dark. Even if it is there like that."

Everyone in the area knows that a primary cause of the lagoon's ongoing contamination is the informal recycling of discarded electrical and electronic equipment in Agbogbloshie. E-waste processing at Agbogbloshie is part of the flourishing informal sector in a rapidly urbanizing Accra (Grant and Oteng-Ababio 2012; Oteng-Ababio 2012) that serves and supports the livelihoods of 4 million daily. As mentioned in the Introduction, Agbogbloshie began as a general scrap yard specializing in vehicle repair, automobile spare parts trading, welding, and tire servicing. In the early 1990s, the city authorities, in an attempt to decongest the central business district of Accra, relocated hawkers and Accra's yam market to the edge of the Korle Lagoon. This relocation made it possible for the scrap market to build from the ground up, offering a space for various services such as vehicle repair, spare parts trading, welding, auto mechanics, and tire servicing that were crucial to the operation of the yam trucks (Grant 2006). The development of the scrap yard was further facilitated by the inflows of economic migrants, primarily from Ghana's northern regions, an area which suffered diminished agricultural opportunities and compounding intertribal conflicts at the time, as discussed in Chapter 2. In the absence of job opportunities in the formal sector, precarious options in the informal sector, and rising housing costs and rent in Accra for the urban poor, most of these migrant laborers found residency in the Agbogbloshie/Old Fadama area. Importantly, in an environment of rapid growth in informal sector activities around the broader Agbogbloshie area, e-waste processing quickly became a part of the Agbogbloshie scrap market that started around vehicle repair. Currently, the Agbogbloshie scrap market plays a central role in Accra's industrial activity, having become a booming informal economic hub; and, much like other regions of greater Accra, it has since the early 1980s become an urban neoliberal "development" project and, more recently, a space of contentious "green" intervention (Little 2016; Little and Akese 2019). Following Ferguson (2005), so-called cleanup and toxic waste mitigation in the Korle Lagoon is currently informed by a neoliberal governance and management approach that has spread across Africa, an approach that ultimately favors less state involvement and more NGO capacity-building and problem-solving.[9]

As a primary node of a vibrant scrap metal trade industry that generates between $200 and $300 million in revenues annually in Ghana (Daum, Stoler, and Grant 2017), the Korle Lagoon is one of Ghana's busiest industrial areas. For example, sitting right at the banks of the lagoon, the settlements flood during the peak rain season of June–July. Furthermore, with blame for the supposed death of the lagoon hanging over the settlements' shoulder, many people in Agbogbloshie and Old Fadama live in constant threat of eviction from the city authorities, whose default policy for dealing with urban floods is evictions of communities living close to waterways. For example, in the wake of a twin flood and fire disaster that claimed the lives of over 150 people in Accra on June 3, 2015 (see Figure 3.5), parts of Agbogbloshie and Old Fadama were violently demolished (Lepawsky and Akese 2015). In the process, over 20,000 residents, mostly women and children, were rendered homeless and had to struggle to find shelter from harsh summer rainfall. A resident in Sabon Zongo, a predominantly Hausa-speaking neighborhood adjacent to Agbogbloshie, explained to me that many people blame the explosion on a person who carelessly threw a cigarette near the gas pumps; but inspectors of the disaster never determined that this was the cause of the explosion.

---

[9] As Ferguson (2005, 379) puts it,

According to the mythology of neoliberal globalization, the reforms of Africa's "structural adjustment" were supposed to roll back oppressive and overbearing states and to liberate a newly vital "civil society". The outcome was to be a new sort of "governance" that would be both more democratic and more efficient. Instead, the best scholarship on recent African politics suggests that the "rolling back" of the state provoked or exacerbated a far-reaching political crisis. As more and more of the functions of the state were "outsourced" to nongovernmental organizations (NGOs), state capacity deteriorated rapidly.

Figure 3.5. The petro station fire, June 3, 2015.
Source: http://myacadaxtra.blogspot.com/2015/06/over-70-killed-in-ghana-petrol-station.html.

It seems unclear how the AMA will respond to another flood and whether or not these structures will stay intact or be bulldozed in the aftermath of a future flood. As Daum, Stoler, and Grant (2017) point out, immediately following the 2015 flood, the AMA had originally planned to bulldoze 100 meters from the lagoon but, after negotiations with community leaders in Old Fadama and Agbogbloshie, limited the bulldozing to 50 meters. Despite these negotiations, conflicts between community members and the AMA erupted when bulldozing occurred beyond 100 meters from the lagoon. As local research on the situation notes, "a fear of more extreme backlash from the cadre of informal e-waste workers discouraged the government from bulldozing even further into the community. Despite the absence of additional bulldozing operations in 2016, the municipal authority [the AMA] has given notice of future demolitions" (Daum, Stoler, and Grant 2017, 2).

This so-called backlash is projected to be coming from the poorest of the poor in Accra. Those struggling most for survival in Accra's urban margins are deemed "backlashing" people, with exaggerated impulses stemming from their "illegal" status as landless squatters.[10] But, more importantly,

---

[10] For an excellent ethnographic perspective on what has been more accurately termed "il/legality," see Nordstrom (2007).

this "backlashing" is embodied. During fieldwork in 2015, Ibrahim showed me marks on his arms and back from aggressive evictions in 2015 in Old Fadama that resulted in some scuffs between AMA soldiers and Old Fadama residents. But he and other Dagomba are certainly not aggressively opposing the state in favor of better e-waste labor conditions. Instead, securing a place to sustain a life as Dagomba in Ghana's urban margins is what drives his fight to have a right to stay in Old Fadama. Moreover, there are cultural events practiced in Old Fadama, practices that tell a story that goes beyond simply displaced informal workers from northern Ghana "squatting" in this industrial area of Accra. For example, important Dagomba rituals are carried out in this area, like naming ceremonies for newborns in the community. But the state's approach to flood control in the area seems to minimize this deeper cultural tie to Old Fadama. Instead, in this cultural space, floods and fires have far more than biophysical effects and consequences. Being "vulnerable" to flooding and petro explosions, for example, involves embracing a more complex understanding of the social and material realities of urban informality itself. As Amoako and Inkoom (2018, 2919) insist, we ought not

> ignore the importance of biophysical factors such as rainfall, overflow of rivers and lagoons, and high tides as important in triggering flood events, but to unearth the complexities of ways in which socio-material and political processes of informal urbanisation in urban areas may shape flood vulnerability of residents. These concerns bring to the fore the argument for a re-look at flood vulnerability research and intervention approaches to consider urban informality as a key ontological and epistemological lens.

The 2015 petro station fire also crept up unexpectedly in a conversation I had with a friend, who also happens to be an educated elite in Ghana. During a visit with a friend who lives near the University of Ghana-Legon, we discussed the social and environmental issues going on in and around Agbogbloshie and Old Fadama. The overwhelming sadness of the flood and fire, it seemed, was coupled with critique of Dagomba continuing to live in such a "high-risk" area, which they know is dangerous.

"You heard about the fire near Nkrumah Circle, right?" Alice asked as she arranged the table for a late lunch.

"Yes, it is so sad! I walked by the site the other day and it was all boarded up. I also saw some of the pictures in the news report." I replied.

"The flooding was so bad this year and they all got caught in it. Those people who live there know it is dangerous. They know. There is no place for them to go, but they know they can't be in that flood area," she commented as she folded napkins in preparation for our messy feast of *fufu*.[11]

"What are they to do though?" I replied, feeling for the first time a feeling of advocacy for a group of Ghanaians that Alice, and many other educated elites in her social network, had little interaction with.

"I feel that they need to be out of there. The government has failed to help them, so they need to help themselves. This is always the problem in Ghana. You know, they don't own that land there, but they want to and that is a problem," she added, noticing that we were playfully agreeing to disagree. "But the fire was so, so bad. So many lives lost. I pray for them. For that we pray."

Alice and I might have wandered into a discussion and debate over the reality of Dagomba rights to the city,[12] but the food was ready and it was time to eat. As we sat eating our delicious meal, my own internal interrogation was hard to ignore. "What if at the heart of the struggle for Ibrahim and his fellow workers is a more chronic struggle over land, yet another struggle of the landless?" I asked myself. The purported "toxic" environment is equally an urban environment of ongoing dispossession, infrastructural violence, and profound asymmetries of sociopolitical power.

Of Landlessness, Dispossession, and Violent Obsolescence

Recent research on the community of Old Fadama has argued that a major point of contention in this urban slum is the fact that the AMA sees a strong distinction between landowner and resident. In an interview with one of the AMA's senior planning officers, a recent report notes that an officer stated the following:

[11] *Fufu* (or *fufuo*) is a staple food common in many West African countries. In Ghana and Nigeria, it is traditionally made by mixing and pounding equal portions of cassava and green plantain flour thoroughly with water.

[12] Inspired by the classic works of Henri Lefebvre (1970, 1991, 2003), many have since engaged and revised the "right to the city" discussion (see Harvey 1973, 2008, 2009, 2012; Mitchell 2003).

We will recognize them as residents of the city, but we will not recognize them as land owners. . . . I don't know if anybody will recognize them as land owners, but their interest [is] as stakeholders [and] as residents of the city, and [with] the current houses they are building. . . . I know in future that if we get adequate funding and if the project comes on stream . . . we should be able to build different types of houses that can be accessed by low-income people so that everybody can be catered for, so there are many different packages and models for that area. . . . [A]s for eviction, it's been out of the question for some time. The relocation is on board and after the relocation there will be fertilisation and upgrading of the area. (Stacey and Lund 2016, 609)

Considering the overwhelming presence of displacement, eviction, and demolition threats, my ethnographic experience in Old Fadama and Agbogbloshie has made me more sensitive to the actual life politics of the landless and the lived experiences of explicit dispossession. While this will be explored further in Chapter 6, I have become especially attuned to the fundamental ways in which land rights figure in efforts to effectively do righteous interventions in postcolonial Ghana. Access to land and being granted land by Chief Yoo Naa has deep significance for Ibrahim. For Ibrahim, being granted a plot means that he has been given a chance to start a business, but the capital needed to start something in Savelugu leaves him with little choice but to migrate to Agbogbloshie to continue with copper wire burning labor, which he knows is directly impacting his health. He knows bodily contamination comes with the cash he can make from making a trip to Agbogbloshie. He also knows that ongoing increases in rent extractions in Old Fadama, where he stays during his labor migrations to Agbogbloshie, make saving money to bring north even more difficult. These difficulties persist for many e-waste migrants, despite the efforts of organizations like People's Dialogue and Shack/Slum Dwellers International. As Stacey and Lund (2016, 607) point out, involvement of these organizations "helps define Old Fadama as part of a global [slum] problem, and contributes to the global profile of these international NGOs as facilitators of urban development with a human face. At the same time, it is acknowledged that the production of such knowledge following the involvement of global institutions may not be favorable to marginalized slum dwellers." Instead, urban agency power is often bolstered in such situations, even opening access to untouched revenue streams which don't trickle down to Ibrahim or any other slum dweller. These capital accumulation and related territorial[13] politics and logics are at the heart of AMA management decisions and actions, despite ongoing ambiguities about determining who and who isn't the "responsible" urban authority in Old Fadama and Agbogbloshie.

In August 2018, the AMA was publicly accused of engaging in demolition projects to increase its own revenues (Lartey 2018). This accusation came from Nii Lante Vanderpuye, a member of Ghana's Parliament who represents Odododio, the primary constituency of the Greater Accra area. Nii Lante noted that the AMA had plans to go ahead with dismantling illegal structures in Agbogbloshie to allow for more tomato sellers to practice their trade. The AMA, he added, had plans to destroy the structures despite opposition by "the squatters" themselves. According to Nii Lante, the forceful eviction of residents is a deliberate attempt by the AMA "to increase its revenue gains. What the AMA wants to do is that they want to bully people to take their lands in order to sustain their revenue generation" (Lartey 2018). The legislature eventually took the side of the tomato traders, arguing that this group of workers in Agbogbloshie provides a main source of income for the AMA. As noted in this news report, "The AMA is making big money from them [and because of that] the AMA has to find a place for them to really settle in order to generate maximum revenue from them" (Lartey 2018). Nii Lante, along with the support of other workers and "squatters" in Agbogbloshie, has tried to advise the AMA to reconsider this demolition plan and instead find another land area in Accra to create a separate market for the tomato workers. But despite the efforts of workers, the AMA has taken a sturdy stance, and the agency exhibits this power by continued justification of these urban land management decisions. The AMA has even stated "We'll eject Agbogbloshie squatters despite protest." According to Nii Lante, "they [the AMA] just want to come in and say the land is government land. The people are squatters and so they are going to demolish and they will not reason" (Lartey 2018).

Gilbert Nii Ankrah, the AMA's head of public affairs, has continued to remind those skeptical of the demolition plan because the AMA has given the squatters enough time to vacate. The ongoing message is that these urban squatters just need to get out, and over 1,000 structures need to be leveled,

---

[13] Social scientists tend to distinguish between "place" and "territory." While "place" is space that is transformable, habitable, gendered, and emotive (Gieryn 2000; Davidson et al. 2008), "territory" is mostly understood as a process of appropriating, claiming, and bordering space by particular social groups to contain and sustain geopolitical, sovereignty, and political economic interests (Brighenti 2010; Escobar 2008).

a stance that leaves little room for a conversation of cohabitation. "We have given them ample time to ensure that they move their belongings and failure to do so we will be forced to eject them. They are not supposed to be there, and the assembly is not making any arrangement to relocate them to any other place," argued the AMA's public affairs leader. Despite their obvious "interest" in Agbogbloshie and sources of revenue generation in Accra, the AMA wants to depoliticize the Agbogbloshie demolitions. The chief executive of the AMA, Mohammed Adjei Sowah, has even accused Nii Lante Vanderpuye and other opponents of demolition exercises in Agbogbloshie, of "stoking partisan fire" (Adogla-Bessa 2018), adding that "if decisions are being taken in the supreme interests of this country and people could think that they can take political advantage of it, I consider such people as desperate politicians and they should think about this country and the safety of everyone." Of course, some are deemed safer than others in this precarious urban environment.

What is for sure is that orbiting this contentious urban political discourse is a process of dispossession that copper wire burners regularly cope with and struggle to live with. The AMA, like most managerial bodies mandated to make decisions and take action in the interest of the public and the Ghanaian state, maneuvers in ways that extend and sustain both its "managerial power" (Butler and Athanasiou 2013) and its protection of the capacity to actively dispossess. While the AMA's stated mission is "to build a smart, safe, sustainable and resilient modern city," we need to endorse a more critical optic of dispossession to honestly know and understand the ground-level actions of the AMA and the negative effect these actions have on the lives of displaced copper wire burners. The AMA, in this light, is doing far more than flood control and waste management, even if that continues to be a highly visible form of "smart" urban environmental governance.[14]

Urban rent extraction and negative "slum" settlement management are also going on amidst the AMA's usual charter, which includes a lot of positive projects, goals, and outcomes (Stacey 2019). After all, the AMA is just one of 254 metropolitan, municipal and district assemblies (MMDAs) in Ghana and among 26 MMDAs in the Greater Accra region. Established in 1898, the AMA has changed over the years, in name, size, and jurisdiction. When Ghana returned to constitutional rule in 1993, it derived its legal basis from

---

[14] The smart city movement has deep ties to big tech, and as I explore in Little (2014, 2015), IBM's "Smarter Planet" corporate campaign involves a toxic paradox.

the Local Government Act, 1993, also known as Act 462, which currently has been amended as the Local Governance Act, 2016, known as Act 936, and under Legislative Instrument 2034. Under Section 12(3) of Act 936, the AMA is charged with meeting a cascade of goals that most Ghanaians I have spoken with agree will likely never be met, despite ongoing parliamentary optimism. To be fair, the AMA has a cascade of overwhelming, even unrealistic, mandates to fulfill as an urban development and management agency. The AMA is "responsible for the overall development of the district" of Accra. It is charged, among other things, with formulating and executing "plans, programmes and strategies for the effective mobilization of the resources necessary for the overall development of the district" and to "Promote and support productive activity and social development in the district and remove any obstacles to initiative and development." Residents of Old Fadama and workers in Agbogbloshie have experienced removal and have known what it is like to be an "obstacle" to urban development many times.

There is also at the heart of any critique of the AMA and its flood control decisions the ongoing battle over the control and management of land. Urban land management customs and quarrels spring up mostly as a result of "the complex overlaps between Ghana's customary land administration and the state's partially functional urban planning systems" (Amoako and Inkoom 2018, 2919). In other words, informal settlements like Old Fadama are born out of complex negotiations between indigenous land owners and state-based urban planners. For some, it is these negotiations that ultimately make places like Old Fadama most vulnerable to flood hazards: "the inability of the city's administrative system to mainstream the customary land market, deal with housing supply deficits, provide affordable housing and community infrastructure has contributed directly to the development of informal settlements in flood-prone areas" (Amoako and Inkoom 2018, 2919).[15]

In light of this, we also need to rethink dispossession in Agbogbloshie and Old Fadama and recognize the actual violent terms that play out on the ground. In other words, "dispossession, as a way of separating people from means of survival, is not only a problem of land deprivation but also a problem of subjective and epistemic violence; or, put another way, a problem of discursive and affective appropriation" (Butler and Athanasiou 2013, 26). What unfolds in postcolonial Agbogbloshie and Old Fadama is more than

---

[15] Millar (2018) has also reported on similar flood vulnerability politics in her ethnographic research on scrapyard workers in Rio de Janeiro, Brazil.

erasure in practice. It is more like planned, and even violent, erasure in a burning e-waste landscape already marked by obsolescence and precarity (Stacey 2019). While Ghana has e-waste management legislation in place, with Act 917 (Hazardous and Electronic Waste Control and Management Act)[16] and Legislative Instrument 2250 introduced in 2016, the actual people engaged in e-waste labor experience a livelihood of urban marginality and postcolonial dispossession on a daily basis that current legislation does nothing to correct. Their shadowed lives, in many ways, exist on the edges of a system of "managed" and even "planned" obsolescence that produces and sustains conditions of urban violence and displacement, rather than creating more socially just conditions to actually help Dagomba people live a good life in Old Fadama free of ostracism and social obsolescence.

A concept crafted in the Global North, "planned obsolescence" began to circulate in the high-tech industry in the 1980s, and it has a straightforward consumption logic that looks at innovation and progress in product development through the optic of intentional disposability. As Slade (2006, 3) points out, intentional or deliberate obsolescence came straight out of the American cultural handbook: "Not only did we invent disposable products, ranging from diapers to cameras to contact lenses, but we invented the very concept of disposability itself, as a necessary precursor to our rejection of tradition and our promotion of progress and change."[17] To practice planned obsolescence was to intentionally make electronics that have a short consumer life so that more electronics could be continuously designed, manufactured, and sold on the electronics consumer market. It offered another method of adapting electronics production with "fast-capitalism," "a corrosive 'productive' regime that transforms the conceptual and the relational power of 'society' by subverting fundamental moral claims, social distinctions, and material dispensations" (Holmes 2000, 5). As with any discussion of capitalism itself, geography matters in any critique of planned obsolescence.

---

[16] Act 917 was specifically designed "to provide for the control, management and disposal of hazardous waste, electrical and electronic waste and for related purposes."

[17] Slade (2006, 3–4) adds,

> As American manufacturers learned how to exploit obsolescence, American consumers increasingly accepted it in every aspect of their lives. Actual use of the word 'obsolescence' to describe out-of-date consumer products began to show up in the early twentieth century when modern household appliances replaced older stoves and fireplaces, and steel pots replaced iron ones. But it was the electric starter in automobiles, introduced in 1913, that raised obsolescence to national prominence by rendering all previous cars obsolete.... The earliest phase of product obsolescence, then, is called *technological obsolescence*, or obsolescence due to technological innovation.

Certain territories and "Black ecologies" (Hare 1970; Roane and Hosbey 2019) have become sinks of electronic discard and toxic burden made all the more possible with the "modern cult of 'disposability'" (Hecht 2020). Certain "technocapital sacrifice zones" (Little 2017), in other words, are produced in response to and as a condition of these powerful disposability logics. On display in Agbogbloshie is a situation of "catastrophic convergence" (Parenti 2011) where planned obsolescence is grounded and lived and where it becomes a socio-material force in a planetary system of technocapital ruination, urban displacement, and pyropolitical violence. When we "turn toward embodied lifeworlds" (Lock and Farquhar 2007, 9) to make sense of the lived experience of these forces, we also confront the hard facts of toxic exposures and suffering, which is the central focus of the next chapter.

4

Embodied Burning, E-Waste Epidemiology, and Toxic Postcolonial Corporality

The copper smoke, it hurt my heart. The fire and smoke disturb me. We suffer here. Burning the copper here, we suffer. My body burn. My body burn in here.

— Awata, Agbogbloshie worker

Wahala mi o dira [The one who suffers knows their suffering].

— Dagbani proverb

Agbogbloshie is no doubt a chemical environment. Alongside copper wire fires and modern marvel discard, from junked cars to refrigerators, coffee and fitness machines, printers and photo copiers, and clunky desktop computer monitors, the scrap market landscape contains numerous toxic substances. Among other hazardous substances, lead, mercury, cadmium, polychlorinated biphenyls, and airborne contaminants, including polybrominated diphenyl ethers, pose significant health risks to a wide range of bodies—both human and nonhuman—making a living in this place. Remember that Agbogbloshie has been described as "one of the most polluted places on Earth" (Pure Earth.org), so it is no wonder why this location of toxic metal extraction has been the focus of numerous domestic and global environmental health science studies.[1] As health scientists have noted, e-waste recyclers who burn copper-based electronic cables are exposed to

---

[1] See, for example, Riederer, Adrian, and Kuehr (2013); Caravanos et al. (2011, 2013); Feldt, Fobil, and Wittseipe (2014); Wittsiepe, Fobil, and Till (2015); Asante, Adu-Kumi, and Nakahiro (2011); Asante, Agusa, and Biney (2012).

Burning Matters. Peter C. Little, Oxford University Press. © Oxford University Press 2022.
DOI: 10.1093/oso/9780190934545.003.0005

extreme levels of harmful toxins from the burning of the polyvinyl chlorides in the plastic that insulates these electronic cables. After all, we now know there is a plethora of toxic substances found in modern electronics and appliances. These notable toxins assist with basic electronic functioning. For example, lead and arsenic are used in cathode ray tubes, and lead is a standard solder. Antimony trioxide is used as a primary flame retardant, and polybrominated flame retardants are used in plastic casings, cables, and circuit boards. Selenium is widely used in circuit boards as a power supply rectifier. Cadmium is heavily used in both circuit boards and semiconductors. Chromium, which acts as a corrosion protectant, and cobalt, which provides structure and magnetism, are used in the steel. Finally, mercury is used in switches and housing units. Epidemiological studies among e-waste recyclers have explored exposures to these various toxic substances and have revealed numerous health effects. Additional studies have shown that children in Agbogbloshie are particularly susceptible to toxic heavy metal exposures, focusing on "hand-to-mouth" behavior, which is common among young children, and other routes of exposure including breastfeeding, placental exposure, secondhand exposure to parents' contaminated clothing, and "skin-to-skin" contact. Research has also revealed that children living in or near e-waste recycling facilities illustrate significantly lower weights, heights, and body mass index compared to populations unaffected by pollutants from informal e-waste recycling. The lead in many of these e-waste sites also presents risks for neonatal neurological development (Liu et al. 2011), a situation that illustrates broader concerns about pediatric environmental health risks and their global significance.[2]

These environmental health science findings have expanded interest in global health programs, leading several researchers to conclude that "the health effects of e-waste exposure must become a priority of the international community" (Grant et al. 2013, 358). Additionally, public health scientists have noted that toxic exposure to e-waste recycling is not limited to e-waste workers alone. Some researchers highlight the fact that "Traders and residents within the Agbogbloshie enclave are equally at risk through a range of environmental vectors," calling "for increased public awareness about the effects of human exposure to lead and other toxic elements from e-waste recycling" (Amankwaa, Tsikudo, and Bowman 2017). The truth is

---

[2] According to the World Health Organization, infant mortality rates caused by environmental exposures are 12 times greater in developing countries than in developed countries (Singer 2016, 11).

there are many living bodies navigating this scrapyard landscape, and the burning of e-waste penetrates all of these interacting bodies. Another vital truth is that while there exists an accumulation of environmental health science data and toxicological evidence, the overwhelming embodied realities of urban marginality, slum dwelling, and lived suffering informed by "structural violence" (Farmer 2003) and "slow violence" (Nixon 2011) are usually overlooked or given scant attention in these environmental health science reports. Following Farmer (2003, 50), no honest assessment of environmental health disparities in places like Agbogbloshie can "omit an analysis of structural violence." This analysis involves the role of domestic and global agencies, nongovernmental organizations (NGOs), and growing public–private partnerships involved in imagining, producing, and administering environmental health science and knowledge in Africa, where the global public health industry has accelerated since the 1980s in response to the African AIDS epidemic (Prince and Marshland 2014).

In addition to being connected to a growing neoliberal system of public–private partnerships, attention to the global health dimensions of e-waste recycling labor is fueled by a crisis and alarmist discourse, which is precisely what made "global" health an international matter of concern in Africa in the first place. Across Africa, postcolonial "development" involves a growing network of civil society organizations and international NGOs, many of which anchor their missions to achieving global health outcomes (Prince and Marshland 2014). Moreover, to justify this shift toward "global" health, intervention was strategically couched in crisis and emergency terms that almost automatically set up a moral dilemma. According to Prince (2014, 28), "global health has taken shape in an ideological framework dominated on the one hand by 'emergency,' 'crisis,' and concerns about 'global security' and on the other by humanitarianism, both of which have proven powerful forces for mobilizing resources and action." These same powerful forces seep into and inform health science projects taking root in Agbogbloshie.

As discussed in more detail in Chapter 6, the toxic situation and environmental health challenge in Agbogbloshie have also drawn the attention of a "solutions-based" international NGO, turning Agbogbloshie into a place of speculative NGO intervention to curb e-waste burning. These interventions include environmental health logics and goals that tend to overlook the actual complexities of "suffering" subjectivities (Kleinman, Das, and Lock 1997). In this sense, to more fully comprehend "environmental suffering" (Auyero and Swistun 2009) in Agbogbloshie, we must turn to the "intricacies of detail which color experience and enable us to appreciate the fine grain of improvisation, novelty, and difference that make life so much more than a model or statistic can convey" (Kleinman 1995, 252). We must turn to how toxic exposure and disturbance are actually felt, lived, and embodied. Workers I interviewed not only referred to "internal" conditions of toxicity (e.g., respiratory problems, heart pains, and headaches) but also turned my attention to the more "exterior" and more visible forms of harm and bodily distress—burns and scars—to make sense of their suffering.[3] Copper wire burners, in this way, have a certain experience of toxic embodiment that breaks down and reconfigures demarcations of body and environment in a transformative landscape of pyrocumulous activity.

This chapter engages the social, environmental, and biopolitical boundaries between e-waste labor, toxic embodiment, and violence amidst precarious risk mitigation interventions and logics. It sheds light on the ways in which "techno toxic" (Precarity Lab 2020) livelihoods, environments, and economies are embodied and understood. These local narratives, I argue, expose new themes at the intersection of pyropolitical ecology, discard studies, and critical environmental health. In particular, I argue that attending to how these workers manage and make sense of the health risks of e-waste labor and NGO intervention helps us understand the bodily distress and toxic ecology of informal e-waste labor in Ghana's urban margins. Several critical environmental health questions inform my focus here and might be used to theoretically augment debates on toxic labor and body–environment relations: What are working bodies and subjectivities up to in a toxic postcolonial urban context? What stories do e-waste laboring bodies and subjectivities tell? What boundaries do these stories of toxic embodiment push or break down? Finally, what is the ethical and political role of anthropology in postcolonial toxic struggles, especially where extreme environmental health challenges persist and where solutions-based interventions overlook the complexity and diversity of eco-corporeal relations? How do current e-waste interventions reduce understandings of risk to chemical toxicity instead of the broader political ecology of risk that actually conditions embodied experiences of toxic labor?

---

[3] Burns are in fact a recognized health concern of the World Health Organization (2011).

Emerging E-Waste Epidemiology

Over the years, Agbogbloshie has grown to include mosques, churches, informal football fields, a goat pasture, and gardens producing agricultural goods. With a meat market in walking distance to the scrapyard, livestock graze everywhere at the site and even rest on the mounds of ash generated from the cable burning. During one visit to the site in 2016, one worker explained that "They [the grazing livestock] are always here. They sit on da ash. After we burn, they wait and sit on the ash." I observed this during my daily visits to the site and was later told that when the ash cools down, the livestock like to rest on it and get it on their fur. "They like how it feel," I was told. Recent studies by researchers at the University of Ghana have confirmed that the meat from these livestock is rich in heavy metals and other toxic substances. These same contaminants, especially high levels of lead, zinc, and chromium, as well as flame retardants, are found in the blood of workers tested at the site (Wittsiepe, Fobil, and Till 2015). The little to no protection worn by these workers has been viewed as the cause of these elevated levels of toxins, but the long-term health effects such practices have on these workers have yet to be studied. Also, to date, there is no public health research confirming the risks of consuming contaminated meat and milk from livestock sourced at the site, even as Agbogbloshie is home to an active meat market.

While many sources of risk exist, e-waste burning in Agbogbloshie has no doubt caused detrimental "environmental suffering" (Auyero and Swistun 2009) and a so-called airpocalypse situation (Ghertner 2020). Copper wire burning is considered to be the primary source of air pollution and is the core environmental health risk as determined by internationally recognized epidemiologists and public health experts at Ghana Health Services, the country's leading public health agency. It is well known that these informal e-waste recycling practices not only release toxins into the air but also leave behind a toxic ash on the soil, which can then blow around and become yet another airborne toxin as well as a source of local water contamination. Additionally, pollution generated at Agbogbloshie is blamed for poor air quality in Accra writ large (Daum, Stoler, and Grant 2017), yet many know that this city of over 2 million is also choked by cars, buses, and delivery trucks that significantly contribute to urban air quality. Adding to the toxic confusion— and blame confusion—is the open incineration of rubber tires, which generates massive toxic plumes of smoke in Agbogbloshie (see Figure 4.1).

These tire burning workers, many of whom come from the northern city of Bolgatanga (or Bolga, as workers refer to it) near the Ghana–Burkina Faso border, are *not* the target of e-waste epidemiology or the focus of toxic mitigation interventions. Some I spoke with suggested this is likely because many of the tires are also burned to singe goat hides and smoke meat. Intervening on this pyrocumulous labor, I was told, would interfere with a much needed food supply. In the words of one worker, "That be for food. The AMA not touch dat." According to one reporter investigating the use of tire burning to process meat in the city of Kumasi in the Ashanti Region (Green 2019), the practice of tire burning is linked to the rising cost of liquefied petroleum gases like propane, which are typically used for cooking. As a result, tire burning is a common practice that leads to dangerous exposures to benzene, sulfur dioxide, carbon monoxide, hydrogen chloride, polychlorinated biphenyls, and various volatile organic compounds.

But, like copper wire burners from Savelugu, these tire burners struggle to make a living in Bolga as most of their time there is spent engaging in subsistence farming. In the rainy season they farm millet, maize, guineacorn, groundnuts, rice, beans, and sweet potatoes; and during the dry season

Figure 4.1. Tire burning. Photo by author.

they irrigate their fields to grow onions, tomatoes, and pepper. When in Agbogbloshie, they burn tires, a category of urban discard that accumulates quickly regardless of seasonality and regulation. Ibrahim tells me these workers can make five Ghana cedi a day (about US$1.25 a day).

According to one report on e-waste labor conditions,

> E-waste recycling in Ghana is organized into abundant small and informal enterprises. Recyclers in Accra are mostly from the poor northern part of the country, a region facing chronic food insecurity. E-waste recycling has been found to be a more reliable livelihood strategy, despite extreme environmental health hazards. Most of the people employed in the e-waste recycling sector are between 14 and 40 years of age, work for 10–12 hours per day and produce 108–168 overtime hours per month. However, most of the workers do not have any fixed working hours per day or per week. (Lundgren 2012, 28)

My own ethnographic research has also revealed that most of these workers would rather be in their villages in the north, especially if their livelihoods could be supported by wages equal to those found in Ghana's e-waste recycling sector. For example, according to Alhassan, who has worked in Agbogbloshie since 2014, "Life north is hard. No money. No work there. That be why we come here. Here it is hard. Nobody care for us. Living north with work is good. I wish for that." As explored in this chapter, putting their bodies (and their health) on the toxic front lines is considered the best of the available options for workers who feel and embody the toxic pains of displacement, extreme poverty, and urban marginalization.

The e-waste risk has also been creatively approached by recent environmental epidemiologists who are combing the use of wearable cameras and air samples to more effectively track toxic exposures. These e-waste epidemiologists are turning toward interventions where workers are actually wearing and embodying science, wearing observational and measurement tools to build e-waste toxics knowledge. Recently, researchers from the University Michigan "designed and applied a method for objectively deriving time-activity patterns from wearable camera data and matched images with continuous measurements of personal inhalation exposure to size-specific particulate matter (PM) among workers [at Agbogbloshie]" (Laskaris et al. 2019, 829). Tracking the time-activity patterns of workers, these scientists suggest, not only is a method that complements air pollution sampling but is considered necessary for actually knowing the exposure situations confronted in a particular work environment. In this way, this recent study is stretching—maybe modernizing—the range and content of bio-informatic collection practices and data points. Time-activity data is even becoming a biometric beyond blood, a derivative bioeconomic measure to advance e-waste epidemiological knowledge of the risks Ghana's e-waste workers face every workday. To document this workday activity, this team of researchers created a "visual activity dictionary," a code book for time-lapse images captured by cameras worn by e-waste workers. From these cameras, the researchers collected over 35,000 images over 170 work shifts. The images capture what the researchers call "time-activity patterns," or what workers actually do during the work day. The images collected were classified (i.e., sitting, walking, burning cables, smoking, eating, etc.) and synthesized with air pollutant exposure data, or PM2.5 (small particulate matter 2.5 micrometers or less in diameter). To collect the PM2.5 concentrations, the researchers had the e-waste workers wear a backpack with a minute-by-minute PM detection device to measure real-time work activity exposures to toxic substances.

Going beyond the extraction of blood, semen, and maternal urine to measure lead and cadmium "body burdens"[4] is a new direction for e-waste epidemiology. What these new biometrics of e-waste labor risk add, researchers argue, is a less intrusive and burdensome method of data extraction: "Wearable camera data . . . eliminate the participant burden and literacy requirements associated with workers keeping active time-activity diaries with 5 or 15-min resolutions; such high-resolution diaries may be required in job settings with frequent task changes and acute exposures" (Laskaris et al. 2019, 832). But what exactly do these new biometrics and "time-activity data" tell us about e-waste environmental health risk as it is socially and economically experienced? The answer is, very little. This is perhaps why the researchers note that future research methods "should involve an iterative process between workers, local leaders, and multidisciplinary teams, including engineers, exposure experts, epidemiologists, and *social scientists*" (Laskaris et al. 2019, 840, emphasis added). The assumption of e-waste epidemiologists in Ghana is that this population is made up of bodies navigating chemical exposures, but if social science engagement is actually taken seriously, as some have suggested (Little 2019; Little and Akese 2019; Akese

---

[4] This term is almost universal in environmental epidemiology discourse and refers to the burden caused by bodily accumulation and absorption of chemical substances.

and Little 2018), the biological frameworks guiding e-waste epidemiologists working in Ghana will surely need to be upgraded and informed by more critical social science engagements with "the body." So far, there has been no direct biopolitical critique of e-waste in Ghana and China, even despite the fact that all e-waste studies to date deal with or involve a focus on working populations, laborers situated in states with biosocial power, and bodies (as biodata) being studied and governed by epidemiological sciences and global health research teams and science partnerships.

This recent e-waste occupational health tracking study in Agbogbloshie illustrates a broader effort to build a more evidence-based global health perspective on e-waste and its urban environmental health dimensions. As it turns out, this research project is following the same trajectory of other global health programs and trends. As Tichenor (2016, 105, emphasis added) points out, "Current forms of global health funding require *tightly regulated evidence*, proving to the international aid community that their aid has real impact on the local situation, to justify supporting health programs in the Global South." Producing evidence of urban environmental hazards and risks is also often given special priority for interventions in "spaces of uncertainty" (Zeiderman 2015).

Contextualizing Fiery Postcolonial Risk

It is difficult to fully comprehend e-waste health politics in Agbogbloshie without full consideration of colonial, postcolonial, neocolonial, and decolonial contextualization in Ghana more generally. Ghana became an independent nation-state in 1957, a break from the British led by Kwame Nkrumah that sparked a pan-African independence movement. The unevenness of global economic geography did not change dramatically as a result of this independence movement. One reason was that many British companies stayed put after 1957 and, in fact, had a monopoly on Ghana's mining industry. Today, China is the new foreign power, expanding mining operations and financing development projects throughout the country. Africa isn't poor, the saying goes; instead, its wealth is stolen. As Nkrumah himself put it in 1965 in his book *Neo-Colonialism: The Last Stage of Imperialism*, "Africa is a paradox which illustrates and highlights neo-colonialism. Her earth is rich, yet the products that come from above and below her soil continue to enrich, not Africans predominantly, but groups and individuals who operate to Africa's impoverishment." Today, the toxic disposal and recovery of count-less digital devices and electronic appliances in Africa—from Senegal to Ghana to Tanzania—have brought urban poverty and pollution together in new ways.

While recent scholarship on Agbogbloshie highlights how "e-waste is a multidimensional phenomenon" that calls for a more "sustainable trajec-tory for e-waste policy" (Daum, Stoler, and Grant 2017), there is no crit-ical engagement with these issues through the lens of postcolonial critique and decolonization. This omission may not be intentional, but confronting the enduring history, logics, and power of coloniality matters if we truly wish to develop a deeper understanding of, or even imagine, a postcolonial environmental health politics in Ghana. How we make sense of e-waste health and injustice in Agbogbloshie calls for this critical engagement. It isn't a detour from the e-waste debate but rather central to efforts to reckon with the mix of conditions and projects leading to ongoing biocultural and eco-social struggle. As Fanon put it in his *The Wretched of the Earth*, life under decolonization is all about enduring delinquencies and forms of semiotic struggle that rarely, if ever, escape violent rupture: "National liberation, national reawakening, restoration of the nation to the people or Commonwealth, whatever the name used, whatever the latest expres-sion, decolonization is always a violent event" (Fanon 1963, 1). The "event," as postcolonial and phenomenological social sciences have long argued, is materially embodied and therefore thrives in actual bodies bound to locations and conditioned by histories of struggle. Making sense of con-taminated postcolonial bodies in Agbogbloshie is a matter of figuring toxicity within situated histories and forms of bodily and environmental violence (Peluso and Watts 2001). In this way, unraveling the racial pol-itics of e-waste struggle in Ghana confronts a Fanonian calling. Using Fanon to make sense of e-waste suffering in postcolonial Ghana is to think about e-waste as a subject matter with many echoes. As Mbembe (2017, 170) notes, "Fanon's great call for opening up of the world will inevitably find many echoes. We can, in fact, see this in the organization of new forms of struggle—cellular, horizontal, lateral—appropriate for the digital age, which are emerging in the four corners of the world." But while e-waste (as simply material) transcends the boundaries of the nation-state, digital rubbish anywhere confronts structural racism and injustice, the defining features of the "burning house," as James Baldwin ([1962] 1993) would have it.

The struggle over environmental health and toxic injustice in Agbogbloshie involves a history of bodily and discursive violence that persists in a postcolonial Ghana marked by ongoing reanimations of racialization. According to Pierre (2013, xii) "As a postcolony, Ghana's contemporary realities are connected to an interlinked set of practices, experiences, and belief systems— a specific predicament of the long history of European empire making." Amid this imperialistic state-making or statecraft, Pierre (2013) argues, is postcolonial "racecraft": "A modern, postcolonial space is invariably a racialized one; it is a space where racial and cultural logics continue to be constituted and reconstituted in the images, institutions, and relationships of the structuring colonial moment" (p. xii). Agbogbloshie, in this way, is a place where colonization, contamination, and racialization are contentiously entwined by new and emerging practices of neoliberal risk management and NGO intervention, practices experienced largely by the dispossessed and "embedded bodies" (Lock 2015; Niewöhner 2011) of the Global South.

Making sense of e-waste or any other discarded "stuff that is being governed" (Gregson and Crang 2010, 1027) calls attention to certain dynamics of labor, trade, and enduring forms of exploitation in postcolonial Ghana. As Pickren (2014) might suggest, building a critical environmental health approach to e-waste risk in Agbogbloshie needs to "account for the *absences and/or ambiguities* of the law, particularly around exemptions and exclusions for certain materials as well as trade and labor conditions" (p. 31, emphasis added). Moreover, economic sectors emerging or being reorganized in Ghana to aid "economic" development—and the Agbogbloshie scrap market is a major player in Accra's urban industrial sector—face hard postcolonial challenges. In the postcolony, as Mbembe reminds us, "hardly any sector . . . is free of corruption and venality" (2001, 85), even if the original turn to NGOs in Africa in the 1980s and 1990s was considered a possible remedy to state-based corruption and control of health and human services (Prince and Marshland 2014).

One "sector" that has rapidly emerged and been sustained in Ghana at least since 2005 is the scrap metal sector, a sector that involves e-waste extraction practices that have long drawn the attention of global environmental health scientists. Human exposure to e-waste is very complex, with multiple routes and lengths of exposure depending on the activity in which the person is engaged. As Grant et al. (2013) suggest, e-waste exposures can be sourced or categorized into three sectors: informal recycling, formal recycling, and exposure to hazardous e-waste compounds left over from recycling practices.

Formal e-waste recycling centers can possibly help to solve this problem and further protect workers from the health effects of toxic exposures; however, these facilities can be expensive to build and sustain, and so are mostly uncommon in the poorest e-waste hubs around the world. In fact, even while Agbogbloshie, has become an experimental space for "formalized" e-waste recycling, such projects can be tainted by chronic operational and capital investment uncertainties (see Chapter 6). But, in general, the lack of these formalized and "greener" recycling facilities for residents living within a certain distance of informal recycling spaces puts these people at greater risk from food, water, and soil contaminated by toxic e-waste recycling activities. As research has found, the most common routes of exposure in these areas are inhalation, ingestion, and dermal contact (Grant et al. 2013; Little 2016). In China, for example, where e-waste studies have a longer track record, 165 studies have been conducted to illustrate the correlation between e-waste exposures and physical and mental health, as well as neurological developmental impacts. Of these 165 studies, 23 reported associations with physical and mental health, while 16 studies alone illustrated strong associations between e-waste exposure and physical health, showing impacts on thyroid function, reproductive health, lung function and growth, and adverse changes in cellular function (Grant et al. 2013).

According to a 2016 national health report produced by Ghana Health Service (GHS),[5] fire is actually listed as the seventh most prevalent public health hazard, after hazards like Ebola, cholera, meningitis, floods, terrorism, and yellow fever. Interestingly, the same report breaks down these public hazards as either biological, physical, societal, or technological (Ghana Health Service 2017). Fire is listed under "technological" hazards, yet the report never mentions e-waste burning or incineration as a potential hazard. In fact, chemical risk, which would include toxic substance exposure, is list seventeenth, even though e-waste fires surely emit hazardous toxins. The report does recognize the impact of situations where hazards are combined (e.g., flooding and cholera, drought and meningitis), but fires and toxic substance exposures are never mentioned. The intervening NGOs and non-state actors attend directly to this synthesized environmental health risk, but Ghana's

---

[5] The GHS is the largest public sector agency under the Ministry of Health and is "responsible for ensuring that every Ghanaian has access to healthcare services when they need it. The GHS is mandated through its directorates and health facilities to provide preventive, promotive, rehabilitative and curative health services at all levels, to ensure continuous contact and a seamless referral system that enables continuity of health services to every person" (https://www.moh.gov.gh/ghana-health-service/).

leading public health agency does not directly combine these hazards, a situation that further exposes the power of NGOs in emerging global environmental health projects focused on e-waste.

Even while the electronics recycling sector has created some jobs for the working poor around the globe, the toxic waste this sector generates is hazardous to social and environmental systems. Knowing this, developed countries have adopted appropriate management systems to help divert and transport e-waste out of electronics-producing countries. This waste is then transported to developing countries like Ghana, which has an informal sector to take in this waste and help recycle it to turn a profit. However, Ghana lacks the proper legislation and infrastructure to properly handle these flows of e-waste disposal and the global system of digital "donations" (Fuhriman 2008). From 40% to 60% of e-waste is recycled, and 95% of this waste is recycled informally or in a labor sector where environmental and human safety concerns are not a priority or not under any formal management. Many Ghanaians would agree that replacing the informal sector with a formal sector is impractical because the majority of the country relies on informal labor. As Oteng-Ababioa, Amankwaa, and Chama (2014) have noted, reliance on informal labor has made Ghana's urban e-waste recycling sector what it is today, a complex economy with complex socio-political arrangements and forms of negotiation. In general, waste picking has cemented itself as part of the urban economy supplying employment for many impoverished people. The collection of e-waste can be difficult and expensive to ensure that it gets to the proper treatment facilities, and this is where researchers observe the most overlap between the formal and informal markets. Ghana receives over 215,000 tons of waste from the developed world every year, and the informal recycling sector experiences little to no regulation. Some suggest that the ideal model for e-waste recycling in Ghana is to integrate the informal and formal sectors to combine the interests of all groups participating in the e-waste economy, an economy that is responsible for roughly 95% of the recycled waste in Ghana (Oteng-Ababioa, Amankwaa, and Chama 2014). What complicates matters is that one of the primary means for managing waste in Ghana, and especially in the Agbogbloshie and Old Fadama areas, is the simple solution of open incineration. Burning surely reduces waste volumes faster than any other process, yet it is also understood that fire poses significant public health risks to Ghana's citizens and no doubt contributes to the urban air pollution crisis.

E-Waste Burning as Embodied Burning

Unraveling the toxic labor experience in Agbogbloshie calls attention to numerous critical environmental health questions: What links bodies, subjectivities, illness, and toxic fires in an urban market context? How are environmental illnesses and scrap metal economies entangled, and how is that entanglement embodied, narrated, and made legible? In what ways does violence in the urban margins reshape environmental health experience? What do health interventions attend to, and what do they overlook or miss? What possible moral violence does this omission activate?

As I have argued elsewhere (Little 2019), making sense of toxic corporality and permeability in Agbogbloshie calls for going "beyond the body proper," an anthropological perspective on the body that encourages discussion of "a lively carnality suffused with words, images, senses, desires, and powers" (Farquhar and Lock 2007, 15). In this way, if we are to make sense of the biopolitical nature of the environmental risks of e-waste labor, bodily experience itself provides a critical starting point for developing an understanding of larger body–environment relations. As Dietrich simply puts it, "While environmental change can have serious impact on ecosystems, and thus on human society, human bodies remain a key site of experience of pollution, and institutions like the state have profound impact on the control of both people and the environment" (2014, 10). In Agbogbloshie, or anywhere, we confront not only biologized or cultured bodies but materially "embedded bodies" (Lock 2015). Epigenetic embeddedness is also important to consider. Especially in our chemically inundated epigenetic age, all ethnographies of toxicity are challenged to grapple with the "inextricable multiplicities among material bodies and environments past and present: historical/social/political variables, and subjectivities" (Lock 2015, 153).[6] These ethnographies also tend to call attention to the growing convergence of *bio*, *eco*, and even the *geo* ontological (Povinelli 2018, 2019) politics of toxic and wasted environments. From this perspective, land, body, economy, and toxic substances are all entwined and comingling in e-waste recycling hubs like Agbogbloshie, a situation which invokes a critical approach to body–environment boundaries

---

[6] Exploring the force of epigenetics among reproductive toxicologists in China, Lamoreaux (2016) has advanced anthropological discussion of epigenetics by posing the question "what if the environment *is* a person?" (my emphasis), meaning the individual can stand in for both *being* and *representing* the toxic environment. The boundary between body and environment, the person and their surroundings, in this case, collapses, leading to a critical rereading of the individual as the environment.

and draws attention to the "leaky" nature of these boundaries and how this material leakage influences understandings of waste landscapes (Reno 2016).[7]

In an effort to navigate these porous corporeal boundaries, it is important to highlight the ways in which e-waste workers refer to not only "internal" conditions of toxicity (e.g., chest and heart pains) but also the more visible "exterior" conditions of their bodies (e.g., dermal ailments and showing physical exhaustion) to make sense of and make legible their own toxic corporality. This involves asking how we come to know the bodily experience of toxicity and what in fact are the observable boundaries between bodies and environments, between object and subject, for those making a living in Agbogbloshie? It also involves asking ourselves how or to what extent ethnography can indeed expose dynamics of postcolonial corporality. How does focusing on the lived experience and identity of those working and living at the front lines of toxic e-waste recycling in Agbogbloshie inform our understandings of pyropolitical embodiment? To navigate these important questions, I engage ethnographic findings highlighting the lived experience of embedded bodies and lived subjectivities[8] dealing with a political ecology of e-waste risk composed of relations of power at the intersection of bodies, toxins, and intervention logics and practices.

As previously mentioned, on my first visit to Agbogbloshie in July 2015, I met Ibrahim, a young e-waste worker from Savelugu in the Dagbon region of northern Ghana. He was 20 when we first met, and when I returned for fieldwork in the summer of 2016, he was married and had a week-old baby. Ibrahim is of the Dagomba tribe, like many of his co-workers in Agbogbloshie and family living in the adjacent settlement of Old Fadama. Like other young men from Savelugu, he travels to Agbogbloshie for 3 months at a time, two to three times a year (see Chapter 2). Beyond the difficulties of making a living as a young migrant laborer and new father with little formal education, toxic exposures and health problems are additional stressors he lives with. Ibrahim

---

[7] As Reno rightfully suggests, "we tend to imagine waste as an abstract form of social/material, something static and undead, when in reality it is unavoidably entangled with multiple life forms and forms of life" (2014, 20).

[8] Engagements with "subjectivity" are often in close dialogue with Foucault's (1988) "technologies of the self" perspective and the insights of Agamben (1998), who sees the sovereign subject as one who binds to their "own identity and consciousness, and at the same time, to an external power" (p. 5). Here, I approach subjectivity as "the means of shaping sensibility" (Biehl, Good, and Kleinman 2007, 14).

knows the ins and outs of local scrap business, and he also knows the bodily distress that comes from copper supply chain labor.

Ibrahim has reported to me several times during visits that his health is not good. "My head is hot" or "My waist is in pain," he would share. He also mentioned more than once his struggle with what he calls "chesty cough" and that he has difficulty sleeping. He tells me he struggles with chronic headaches and full-body exhaustion. Epidemiologists from abroad and from GHS, Ghana's primary public health agency, have collected his blood and urine for their epidemiological studies. This biodata has been used as evidence in articles published in credible science journals, but Ibrahim has no idea what the results are or what researchers are doing with his blood and urine and why routine (annual) collection of this bodily matter of e-waste workers is occurring. He has no laboratory data on the levels of lead and cadmium in his body, but he does know that e-waste labor is not good for his health. He knows this work burns his lungs and punctures and scorches his skin. When I first met Ibrahim, I asked about his health experience. Without hesitation, he pointed to his burns and scars to show me the bodily impacts of his labor. For him, these forms of bodily harm communicate everything you need to know about the harsh reality of working in Agbogbloshie. He made this particularly clear to me during an interview in 2016, shortly after his first son was born: "Look at dis. Dis be dat. The e-waste. It burn my body. We burn our body here. We hurt. We know it is no good. But how do we live? Life in north is hard. We need this here. My body is bad. Bad body doing this work." For Ibrahim, navigating a life in one of the "most polluted places on Earth" is considered the best of two options: struggle to farm in the savannahs of the Northern Region and make minimal wages in a depressed agricultural market or tough it out doing toxic scrap metal extraction labor in Agbogbloshie to make slightly better, yet still working-poor, wages. For Ibrahim, and many others, coping with the toxicity of e-waste labor is a matter of *economic* survival, perhaps even a matter of bioeconomic vitality and resilience. Physical harm and injury are, of course, felt concerns; but they are also concerns overshadowed by basic needs (Table 4.1).

Agbogbloshie's informal economy also highlights deeper social logics of wealth distribution. The practice of distribution, as Ferguson (2015, 90) notes, "is a crucial social activity that is constitutive of the social (and not only the economic) order. Accordingly, we need to pay attention to the idea of distribution as a necessary and valuable social activity." When Ibrahim talks

Table 4.1. Selected E-Waste Illness Narratives

"Pain in here [points to ribs and heart]. The stomach hurt. Chop [food], small waste landing."

"My chest hurts. Hard to sleep. Eyes be hurt, be burning."

"Sometimes my body burns. The copper smoke disturbs me. At night I wake up two or three times because of heat. My son Martin also sick. He got malaria. He is disturbed by the smoke and is stuffed up all the time."

"My body is no good. Fire work is hard work for here."

"The fire and smoke disturbs me. The chest hurts. I not go to Korle Bu [hospital] for checkup."

"No good breathing. Plus my stomach hurt from small chop. Nobody come to help us with health. They test only the senior scrap workers here."

"The smoke hurts my lungs. It disturb my lungs."

"The smoke disturbs me. The fire heat hurt. The fire makes the head hot. I went to Korle Bu one month ago to check my health. No blood test. Nobody comes here to test you."

"I have trouble sleeping. Chest hurts. The heat bother me. I have medicine for chest pain. The water is hot [water used to cool down hot copper] and hurts the skin. It get in my eyes and burn."

"My chest burn. I take medicine for cough. I also get cut and burned."

---

to me about making money and going back to his village of Savelugu, he is describing his distributive life plan, a life plan that involves his commitment to his birthplace community and family. His productive labor is more than his alone. In other words, "*being* someone [continues] to imply *belonging* to someone" (Ferguson 2015, 148, emphasis in original). Ibrahim carries with him the responsibility of distributing his wages to his family living in Old Fadama and Savelugu. He also shares photos of his injuries with me as a way both to stay in touch and report on conditions in Agbogbloshie and to expose environmental health risks that speak directly to the vital politics of permeability and the dermal threats of copper burning.

Surrounded by piles of scrap metal, toxic fumes from open flames, and malaria-carrying mosquitos, Ibrahim experiences risks in Agbogbloshie that go beyond the everyday bioaccumulation of heavy metals. He deals, first and foremost, with the risks of fire and smoke in general and the open incineration of the countless petroleum-based materials scattered throughout the scrapyard. The burns on his body are symbolic of the "new materialism" turn in the anthropology of the body, where the body "is at once subjective and objective, carnal and conscious, observable and legible" (Farquhar and Lock

2007, 11). Toxic risk is made legible through the body, the skin. Self-reported body burns were actually a key finding in a recent e-waste worker study conducted by researchers at the School of Public Health at the University of Ghana (Armah et al. 2019). In this study, which was based on a survey of 260 e-waste and non-e-waste workers, researchers found that in addition to workers reporting eye and breathing problems, "resident e-waste workers were 84% more likely to report skin burns compared with their counterparts who were resident non e-waste workers. Individuals who had attained tertiary education were 98% less likely to report skin burns compared with their counterparts without any formal education" (Armah et al. 2019, 69). While the latter finding likely remains ambiguous across the e-waste worker population, it makes sense that those with more formal education would not opt for or need to work as copper wire burners in Agbogbloshie and, therefore, would likely reduce certain burning-related health outcomes. The fact of the matter is that e-waste workers laboring in Agbogbloshie don't necessarily point to blood lead levels or flame-retardant exposures when surveyed or asked to explain their experience with environmental and occupational health risks. While toxic metals and substances have certainly bioaccumulated in their bodies, their actual laboring bodies are reflective of capitalism's "nocturnal economy," (Mbembe 2017, 130), a violent economic system that "has always depended on *racial subsidies* to exploit the planet's resources" (p. 179).

During my interviews with copper wire burners, I asked them about the health challenges they face when doing this kind of work. Most workers would begin their response to my question by directing my attention to bodily biomarkers of e-waste labor. "Look what it do to me," Jacob explained while showing me an infected wound on his leg. "This happens to us. We suffer like this. We suffer on our hands, feet, eyes, legs. All over we suffer here. The burning hurts us here."

In this way, e-waste workers' arms and legs tell an embodied story of e-wastelanding in Ghana, a story of wasteland violence and Black ecological suffering (Hare 1970; Roane and Hosbey 2019). Of the workers I have interviewed since 2015, all have mentioned some form of bodily hurt and suffering. Along with the embodiment of a cocktail of toxic substances emitting from the copper-burning flames, they all have been blasted by the heat and flames of wire burning labor. "Look. Take a picture," Abdulai insisted. He extended his arm in front of me and explained what happened. "This happened last year. I fell on the fire with it. Big pain. So much pain. I had to go to

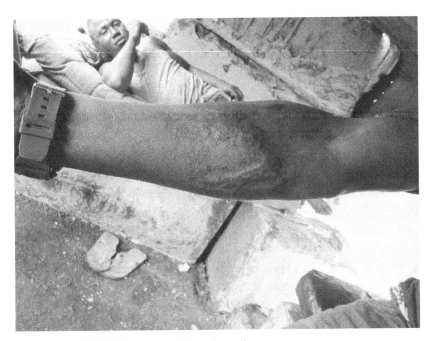

Figure 4.2. Abdulai's burn scar. Photo by author.

hospital for help. I went off to Korle Bu hospital. We have this here. Burning is no good" (Figure 4.2). At the fiery and noxious front lines of e-wastelanding and e-pyropolitics, then, we witness a sort of burning biopolitics.[9]

Beyond occasional cuts caused from transporting scrap to the burn site and the even more common experience of burns to the arms and legs, the bioaccumulation of toxic substances makes workers abnormally tired. One of the most important lessons learned from my time spent with these e-waste workers is the pure exhaustion they experience and endure on a day-to-day basis.

The heat, toxic overload, and long hours put in each day burning copper leave them utterly exhausted. Workers seem to find any chance they can to lay down and escape the sun and heat of copper burning. In my ethnographic experience, this was a front-and-center practice of coping with what is tiresome work in a highly toxic work environment. In the e-waste worker health study of Armah et al. (2019), the survey only captured worker reports of eye

---

[9] A word I jotted down in my fieldnotes reads "biopyropolitics," a term that seems fitting for capturing the vital bodily experience of burning labor.

problems, skin burns, breathing difficulty, and coughing. But with careful participant observation over several months during the course of 4 years, one learns that workers experience everyday exhaustion, toxic heat exhaustion, that is surely linked to the everyday labor environment they embody, navigate, and struggle to cope with.

In Agbogbloshie, there are both exhausted electronics and exhausted bodies. Scenes of resting in Agbogbloshie are rarely the images captured by journalists participating in telling the story of e-waste disaster. Neither are these experiences of bodily exhaustion captured in any environmental health or academic research conducted at the site. As shown in Figure 4.3, Adjiba and her growing baby rest in the shade of the workers' shelter, and she seems curious as to why I would want to know her story. This makes perfect sense because most research conducted in Agbogbloshie completely overlooks women and their toxic exposure experience, even though some studies conclude that women and prenatal exposure risks should be a primary focus of public health interventions (Heacock et al. 2016).

When I returned for fieldwork in the summer of 2017, I interviewed 19 women who work at the site, mostly women who sell food, drinks, and other

Figure 4.3. Adjiba resting. Photo by author.

goods to the men burning for copper. In addition to learning about their migration stories, sources of income, and social ties in Agbogbloshie and Old Fadama, I was interested in the actual experience of mothering in such a toxic environment. Based on ethnographic interviews, I found that women and mothers making a life in Agbogbloshie have a health and "toxic exposure" experience that is too often overshadowed by the hypermasculine e-waste gaze and visual economy dominating representations of Agbogbloshie, a topic I return to in more detail in Chapter 5. But when asked about their health, some women simply said "I am fine," "No problem," or "No issue. I feel good"; but others shared more environmental and occupational health information. Humu, who is from Tolon and sells water to the e-waste workers, said "I am coughing. Most from the weather. My chest come pain. My body come pain." Sirina, who is from Tamale—what some Ghanaians call the "NGO capitol of Ghana"—sells various goods in Agbogbloshie. She is married and has three children, all boys. "I had four, but one died, so I am left with three." Sirina is 27 years old and has been working in Agbogbloshie since 2000. She explained that "Sometimes the smoke enters you." Other women report the pain they experience from carrying goods for sale all day long. For example, I was told "I have a weakness in my body. If I carry too long, I pain" or "You hold the pain, the pain from holding." Another woman, who sells *gala* (hard boiled eggs), said "My health is OK. If I sell fast I can go home and rest. I can get sleep. It is heavy to carry the *gala*." A recent report focusing on persistent organic pollutants in eggs consumed in Agbogbloshie notes that "An adult eating just one egg from a free-range chicken foraging in Agbogbloshie area would exceed the European Food Safety Authority . . . tolerable daily intake . . . for chlorinated dioxins by 220-fold" (Ro 2019).[10] Despite this toxic food chain dilemma, none of the workers I interviewed mentioned *gala* as a source of their health problems. Selling goods and spending your day walking around Agbogbloshie, I was told, is tiring. Rama, who is from Savelugu, shared that "My legs are in pain. I stand a lot to sell the water. It can hurt my legs. I get tired working here."

---

[10] As Lepawsky (2019) warns, this toxic egg article recycles another "easy narrative" of Agbogbloshie, insisting that it is a place haunted by foreign, even "magical," pollution entering the country as electronic discard from Europe.

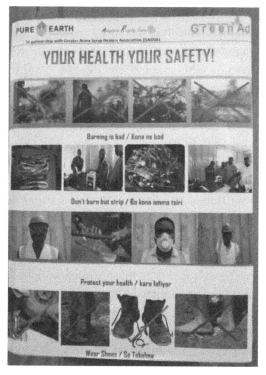

Figure 4.4. Pure Earth health promotion poster. Photo by author.

"Burning Is Bad" and Toxic Demoralization

Local narratives of toxicity and risk in Agbogbloshie exist alongside NGO environmental health interventions and discourses. As noted in the Introduction, in 2014, Agbogbloshie became the site of a "model" e-waste recycling center built to make e-waste recycling work safer and more environmentally friendly. Again, the primary mission: "Eliminate Burning at Agbogbloshie." With support from a variety of government and nongovernment agencies, the new recycling facility, recently named the Agbogbloshie Recycling Center, has a clear risk reduction goal (Figure 4.4). The list of partnering agencies and organizations is long: the European Commission, the United Nations Industrial Development Organization, the Global Alliance for Health and Pollution, the Ministry of Environment, Ghana's Environmental Protection Agency, GHS, the National Youth Authority, Pure Earth.org, Green Advocacy Ghana, and the Greater Accra Scrap

Dealers Association. In short, the new facility aims to reduce the health risks of electronic cable burning—only one of many sources of air pollution in Agbogbloshie—by using automated machines to strip coated cables and wires of various sizes containing copper and other valuable, yet toxic, materials. The e-waste received is primarily large electrical cables—aluminum-based cables—from either Ghana's electrical utility company or neighboring countries like Burkina Faso.

The bundles of wires that the burners collect and burn are much smaller in diameter and are coming from small household electronics, junked cars, buses, and trucks, all urban electronic discard containing copper. These wire bundles range in size and are worth between 8 and 10 Ghana cedi (roughly between US$2.00 and US$2.50), though metal market values directly impact local copper prices. It takes workers about 10–15 minutes to burn off all the plastic insulation. The copper wire is then bagged and sold to scrap metal dealers, usually Nigerians, who then sell the copper for export in Tema, the planned industrial port in the eastern region of Greater Accra. The raw copper is then sent offshore to support growing development and infrastructure projects in China and India.[11] According to market analysis of the global copper trade, China has been a major buyer of copper scrap for many years, but imports began to significantly fall between 2012 and 2016. In 2017, this trend radically changed as the result of a recovering copper price in global markets, spawning a global wave of scrap metal that had been "hoarded" during previous years when prices were lower. China has quickly shifted to this alternative copper source, with imports up 14% to 2.4 million tonnes (bulk weight) between January and August of 2017 (Home 2017). Marginalized agents of this metal recovery and commodity chain capitalism (Jaramillo 2020), Ibrahim and his co-workers are aware of the air pollution they create. They are aware and know because everything is visible, observable, and out in the open to see. As one worker explained, "What we do here everyone can see. You can see our smoke. We are open to see. Other industries are more hidden and so is their pollution. People complain about the smoke being produced here. Our work is very visible. Here you can see exactly how it goes."

---

[11] According to a database managed by Trend Economy.org, the top three export destinations of "copper waste and scrap" from Ghana in 2019 were Korea (US$3.09 million), India (US$1.25 million), and Italy (US$1.07 million).

As seen in Figure 4.4, Pure Earth's health promotion campaign can be read as a "blame the victim" approach, where e-waste workers are told outright that *Kona ne bad* ("Burning Is Bad") and that they should "wear shoes" and care about their health. The campaign aims to offer easy-to-understand solutions to negative practices of informal e-waste labor. While certainly more could be said about the biopolitical dimensions[12] of this e-waste health promotion agenda—particularly the combined forces of self-surveillance and couching recycling formalization within a narrative of good health practices—I want to point out that Pure Earth's approach to formalizing e-waste recycling in Agbogbloshie actually further marginalizes the workers who burn for copper and aluminum. First, they are marginalized by the fact that most of these workers don't have direct access to the facility. Second, fewer than 10 workers in the scrap market—a market with hundreds of workers each day—have been formally trained to use the granulators for stripping the wires. Also, it is well known among all workers in Agbogbloshie that using the granulators slows down the copper extraction process and that it costs money to use the facility since the granulators run on electricity—a constant energy infrastructure problem in Ghana. Finally, the facility primarily processes waste electrical equipment, usually large-diameter cables, coming from Ghana's electrical utility company. Pure Earth's e-waste facility project operates under the "progressive" idea that it can be a win–win for both workers and the environment, but many workers I have interviewed are skeptical of the efficacy and sustainability of the project. No matter how formalized the e-waste recycling sector gets, I am told by one worker, "this facility not good for me or many of us. It be good for someone, but everyone here know that the government want us out. They evict our people [in Old Fadama] who work here. But we do the e-waste recycling. And you see, nobody care for us here."

Some environmental health research in Agbogbloshie has explicitly aimed to employ a "logical framework approach" to provide national, regional, and international stakeholders with "a summary of expert opinion on the most pressing problems arising from e-waste activities at Agbogbloshie, as well as suggested solutions to address these problems" (Cazabon et al. 2017). In one particular study, for example, structured interviews were conducted in the spring of 2015 that used what is known as a "logical framework approach,"

---

[12] For example, one could draw out the links between e-waste health surveillance in Ghana and the racialized biometric politics of surveillance writ large (see Browne 2015).

a scoping exercise used to gauge problems and benefits of e-waste recycling and identify response options. According to the researchers working on this project, the primary goal was to prioritize potential interventions that would address identified problems at Agbogbloshie. Experts concluded that in addition to there being a clear need for further research on the health effects of "primitive" e-waste recycling, e-waste workers should be given "appropriate personal protective equipment." They also concluded that the Ministry of the Environment, Science, Technology and Innovation needs to "revisit" Ghana's Hazardous Waste Bill and that workers themselves be "involved in the planning process of interventions" and "be kept informed of any results." Finally, the report concluded that more education and "sensitization" on hazards related to e-waste was needed for both workers and the general public (Cazabon et al. 2017, 980). Additionally, this "primitive" discourse, aside from being used by NGOs like the Basel Action Network, is replicated by West African environmental health scientists, who report that "In Ghana, the e-waste recyclers use *primitive* methods (mechanical shredding and open burning) to remove plastic insulation from copper cables" (Asante et al. 2016, 145, emphasis added). The point is that multiple actors involved in e-waste health studies and e-waste governance arenas uncritically engage the "primitive" to describe the work being done in places like Agbogbloshie. Use of this term suggests that current informal e-waste labor is backward, negative,[13] and the root cause of environmental contamination, a discursive maneuver that only deepens blame-the-victim logics of urban pollution control.

Toxic Postcolonial Corporality

For scholars of "environmental health justice" (Masuada, Poland, and Baxter 2010; Miller and Wesley 2016),[14] there is no separation of the body and the historic and ongoing forms of discrimination and structural violence marking the body. The same could be said for the relationship between violence—even "slow violence" (Nixon 2011)—and "environment" itself (Peluso and Watts 2001; see also Girdner and Smith 2002). Workers

[13] As Bond (2017) might have it, these forms of copper extraction are illustrative of larger "negative ecologies."

[14] As a community-driven approach to generate scientific knowledge that is relevant to those on the front lines of environmental health risks, environmental health justice aims to "level the playing field in terms of whose knowledge 'counts' in policy decisions that affect disenfranchised communities" (Miller and Wesley 2016, 75).

in Agbogbloshie confront countless social, economic, and environmental forms of distress and violence. Thus, to fully understand the environmental health politics and permeability of toxics in Agbogbloshie, I turn to the term "toxic postcolonial corporality" to honor the fact that Agbogbloshie is also an embodied place and space of postcolonial biopower. It is a term used to navigate the body–toxics–environment relationship in a polluted postcolonial space. In a return to the original coda of colonialism and the so-called productive afterlife of colonialism even in the postcolonial age, Mbembe (2001, 113) reminds us that "colonialism was, to a large extent, a way of disciplining bodies with the aim of making better use of them, docility and productivity going hand in hand." This is useful for thinking about why Ibrahim and his fellow workers migrate from Savelugu to Accra to do "productive" e-waste recycling work in Agbogbloshie in the first place, a migration experience and perspective that was discussed in more detail in Chapter 2. For sure, these migrant laborers can make better wages working in Agbogbloshie since the sale of agricultural goods in northern Ghana is viewed as a losing economic battle, especially for workers like Ibrahim who are newly married, new fathers, and trying to provide for their families. He feels his labor is useless in Savelugu, even though his social capital and status there are praised because his father is a subchief and Ibrahim is himself a drummer in the palace of the head chief of Savelugu, the Yoo Naa. He sees productive value in migrating to Agbogbloshie because it is there that he can make money, money which he brings back to his father and other family members in Savelugu. He, like other copper burners, makes the trip north several times a year. They live a migratory life in general, and a north–south migratory life in particular. "We always back-and-forth," as one burner told me.

As is discussed more directly later (see Chapter 6), there is no shortage of NGO critique to go around, especially NGO involvement in neoliberal Africa writ large (Ferguson 2006). The fact is that Agbogbloshie exhibits a unique case where bodies themselves are seen as the ideal target of global e-waste environmental health science and intervention. Like other global health interventions, those emerging in Agbogbloshie are informed by a larger system of metrics-based humanitarianism (Adams 2016). For example, not only has Pure Earth.org brought an "evidence-based" approach to developing solutions to e-waste in Ghana but this approach tends to favor quantitative data and "metric fixation" (Muller 2018) over critical local health and environmental knowledge(s). This is a common theme in contemporary environmental and global health science. As Walkover rightfully contends,

"Global health metrics, implemented through humanitarian means, promise a lot—better medicines, healthier people—and they produce valuable information and solutions. But when they are assumed to be more valuable than the knowledge that people have about their own lives and conditions, they obscure something much more important" (2016, 176). In other words, despite the importance of pollution control and toxics mitigation efforts, my ethnographic research shows that the most pressing concern for workers is the ability to work and sustain an economic livelihood that can support them and their families. E-waste workers know their bodies are on the toxic front lines, they weigh the options as best they can, but many feel that no matter how hard they try they can't escape the poverty and marginalization they confront in Ghana's postcolonial times and the life they try to live between Savelugu and at the margins of Accra in Old Fadama. Awata makes this point clear to me when he tells me, "Nothing for me. Nothing. That is why I come here. That is why I burn the copper here. Nothing else nowhere." He goes on to add, "I am here to take care of myself. I come for my family in the north. No work in north."

As an ongoing and transformative educational tool and practice (Ingold 2017), e-waste ethnography simultaneously clarifies and complicates what constitutes the lively politics of e-waste. Attending to the embodied e-waste labor experience pushes the boundaries of e-waste materiality, opening up e-waste to its actual constitution and ontological complexity. As Haraway (2020) might see it, "There is no way to touch without inheriting the whole thing." These workers embody the intersection of the postcolonial reuse economy and environment at the same time that they struggle with the toxic body burdens of "recycling" labor itself at the urban margins. They navigate extreme economic and environmental conditions and desire a better life in an African postcolony. It is in this sense that my approach seeks to intentionally tilt the conversation of environmental health struggle in Agbogbloshie toward an understanding of *toxic postcolonial corporality*. As a critical environmental health concept that aims to augment discussion of body–environment relations in a context of postcolonial marginality, it may also be employed to inform the anthropological analysis of permeability in general and the permeable boundaries between electronic discard, toxicity, and postcolonial labor in particular. Much like other work in discard anthropology that calls for going beyond risk and waste itself to advance social theories of waste (Reno 2011, 2014), the term may also expose the need to go "beyond" dominant narratives of toxicity shaping Agbogbloshie truths and falsehoods.[15] The concept encourages a shift in thinking about this vibrant market, a shift that more directly accounts for e-waste embodiment and toxic violence in Ghana's urban margins.

The embodied experience of e-waste workers in Agbogbloshie is more than a situation of xenobiotics, as a recent health study suggests (Armah et al. 2019). A turn to *toxic postcolonial corporality* aims to avoid such bodily reductionism. In this way, the pyropolitical violence explored in Chapter 3, a violence that emerges from lived experiences of erasure, demolition, and urban displacement, calls our attention yet again. We are confronted with a situation where developing a critical environmental health understanding of e-waste labor and life in Agbogbloshie is as much about the everyday bodily distress of working with sharp metal and toxic fire as it is about violence experienced in the urban margins as the direct or indirect result of "the actions and inactions of states and established political actors" (Kilanski and Auyero 2015, 3), including environmental health scientists involved in developing a "global health" approach to e-waste health studies.

In November 2018, I was invited to virtually attend a webinar on "E-waste and Environmental Health" organized by the US National Institutes of Environmental Health Sciences. The webinar explored a case study in China, and the other case study focused specifically on a health study among Agbogbloshie workers. The presenter for the Ghana case study is a researcher at the University of Ghana, and as he explained, his experience doing environmental epidemiology at the site presented many challenges. For example, he explained that "They have had a hard time keeping people in the study. They are always trying to re-recruit people for the study." This researcher went on to add that "We have had some resistance among workers. To keep them registered in the study, we provide malaria treatment. Malaria is so robust, so we treat those that have malaria." Workers have shared with me their struggles with malaria.

One morning during fieldwork in July 2017, I arrived at the worker shelter, where the burners were listening to a local hip-hop High-Life artist ("Stove Boy") on Ibrahim's cell phone. Jacob was sitting, hunched over, while a young boy cleaned his nails. Workers pay one Ghana cedi (roughly 25 cents), usually at the end of their workday when their bodies are coated in toxic ash and dirt. "Kawula [what's up]," I said. "Duso [fine]," he replied softly. "Me catch

---

[15] As Mbembe (2017, 51) might put it, it is a matter of confronting or overcoming the power of falsehoods that is the primary task of e-waste ethnography in Africa.

malaria last week," he added. He said he hadn't eaten much lately and was waiting for his hands to get cleaned before eating what the workers call "concrete" (gari and rice), along with some *wagashi* (fried goat cheese). "I pay 2 cedi for the concrete," he shared.

The fact that e-waste workers' malaria-ill bodies could be a target and justification for environmental health science recruitment smacks of the corporeal politics that seem ongoing in this postcolonial African space. E-waste epidemiology is integrated with malaria surveillance and control. The mission among environmental health scientists working in Agbogbloshie has been to do clean, high-quality environmental epidemiology that meets international environmental health science standards; but, as this researcher notes, enrolled research participants are often noncompliant. In fact, this researcher even noted that it was critical to get buy-in from local chiefs in Agbogbloshie and Old Fadama to conduct more "controlled" studies among the e-waste workers. This is a good practice that meets many of the current ethical standards of community-based participatory research,[16] but that buy-in process has a local political dimension. Dagomba chiefs in Old Fadama and Agbogbloshie are not the ones actually doing toxic e-waste labor. Most copper wire burners, for example, are the bodies on the front lines of pollution. Despite their central role in the scrap metal market economy, these Dagomba workers largely lack sociopolitical power, which is largely attributed to their age. Young boys are taught to respect their elders, especially the chiefs and subchiefs who make up Dagbon political culture. So, negotiation with local chiefs is, in a sense, part of environmental health science practice but never acknowledged as a necessary step or process in generating e-waste worker health knowledge, though it certainly is. Ghana's postcolonial situation is also one made up of chiefs asserting their power, especially in places and spaces where they can, like in the Agbogbloshie scrap metal market. These Dagomba chiefs, it turns out, also inform e-waste health science decisions and negotiations.

Across sub-Saharan Africa today, from Senegal to Ghana to Nigeria to Kenya and Tanzania, waste burning is a common waste management practice. According to a recent World Bank report, "Currently, 69 percent of waste in the Sub-Saharan Africa region is openly dumped, and often burned. Some

---

[16] As Hoover (2017) notes, community-based participatory research methods can improve environmental health science by actively (and meaningfully) involving communities in the development of scientific knowledge. See also Kimura and Kinchy (2019).

24 percent of waste is disposed of in some form of a landfill and about 7 percent of waste is recycled or recovered" (Kaza et al. 2018).[17] This same report goes on to add that e-waste, along with medical and other hazardous wastes, typically makes up a small fraction of municipal solid waste in sub-Saharan Africa. Regarding e-waste in particular, the report notes that its accumulation is directly linked to economic growth and development, "with high-income countries generating five times the volume of e-waste generated by lower middle-income countries" (Kaza et al. 2018, 37). This is a social fact that many Ghanaians know well, as do Nigerians living in Lagos, one of the fastest-growing cities in the world and one of the first places to be the focus of e-waste disaster reporting in Africa (Carney 2006). In fact, it was in Lagos, not Accra, where the story of "mountains" of e-waste and the "dumping of PCs" began to circulate in global news. But regardless of where the e-waste story in Africa begins, it is a story with toxic visualization at its core. To this end, the next chapter explores how Agbogbloshie has become a contentious platform for e-waste visualizations, even a site of productive "ruin porn," to make sense of Africa's e-waste disaster. Yet despite this contentious visual economy and common dystopic prism, there are also alternative techniques of e-waste visualization we can explore. There exist other forms of representation that require more focused attention on e-wastelanding as a cultural, environmental, and place-making process, rather than simply visions of toxic ruination and ecological doom-and-gloom. As will be explored in the next chapter, e-waste visualizations made by e-waste workers themselves can help steer and situate e-waste anthropology more broadly and maybe even guide it down a more just and ethical research and decolonial path.

---

[17] The World Bank has a contentious relationship to toxic waste and pollution issues in Africa. As Ferguson (2006, 70) explains, "On 12 December 1991, Lawrence Summers [former vice president of Development Economics and Chief Economist of the World Bank and former president of Harvard University] sent an internal bank memorandum (later leaked to the press) in which he argued that the export of pollution and toxic waste to the Third World constituted an economically sound, 'world-welfare enhancing trade' that should be actively encouraged by the World Bank."

5

Visualizing Agbogbloshie and Re-envisioning E-Waste Anthropology

How we envision the world matters. But sight is a complex process, a
product of narrative, technology, agency, and matter.

—Gabrielson (2019, 28)

[G]ive yourself over to the circumstances of some other life, hoping to
find yourself taken beyond the limits of your own. These methods are
essential to anthropology's pursuit of humanity as a field of transform-
ative possibility.

—Anand (2019, 6)

Like many others, I first learned about Agbogbloshie through knee-jerk
stories and jarring images circulated on social media, from the environmental reporting of Ghana's Mike Anane to the photojournalism of
Kevin McElvaney, Pieter Hugo, and Ton Toeman. Early in my fieldwork
I contacted Kevin since his portraits of workers in Agbogbloshie seemed
to, in some way, attempt to go beyond e-waste. His visual approach, it
seemed to me, tried to shift the visual narrative of e-waste by intentionally centering the workers themselves, rather than the digital rubbish. His
portraits expose, in some way, the lived personality of e-waste workers. A
photo of a worker standing on top of a discarded computer monitor was
the first image I saw of this mysterious e-wasteland, an image that hints at
the elusive borders of actual and staged rage in this pyropolitical landscape
(Figure 5.1).

Burning Matters. Peter C. Little, Oxford University Press. © Oxford University Press 2022.
DOI: 10.1093/oso/9780190934545.003.0006

Figure 5.1. Portrait of an Agbogbloshie worker. Photo by Kevin McElvaney.

Any critical discussion of the visual economy[1] of e-waste in Agbogbloshie is simultaneously a critical engagement with the methods and ethics of representation, storytelling, and ground-truthing. Much of the circulating e-waste imagery today, especially imagery emerging from the Global South, exposes certain visualization politics and forms of representational distortion. Responding to this call, this chapter explores how, through ethnographic research and community-based participatory research approaches in particular, images make meaning and shape e-waste imaginations. I draw attention to the ways in which images, as techniques of visualization, are deployed by the Agbogbloshie Recycling Center (ARC) and its partnering nongovernmental organizations (NGOs) PureEarth and Green Advocacy Ghana to communicate the alarming environmental health risks of copper wire burning. In attempting to unravel the complexities of this visual narrative, I also address the power and utility of images of e-waste and e-pyropolitics to tell and sometimes distort the stories and realities of and about Agbogbloshie. This chapter explores Agbogbloshie as a key focal point

[1] Attention to "visual economy" in anthropology and cultural studies has led to critical insights into the productive intersection of materiality, representation, and power (see Mitchell 1989; Comaroff and Comaroff 1992; Bennett 1993; Lutz and Collins 1993; Poole 1997; Hutnyk 2004; Sinervo and Hill 2011; Thompson 2015). See also Zuez (2018).

of circulating e-waste visualizations, even a site of productive electronics "ruin porn" in West Africa.[2]

Despite an ongoing and contentious visual economy, there are also alternative techniques of e-waste visualization emerging in Agbogbloshie, especially worker-based representations that require more focused attention on e-wastelanding and e-ruination as processes involving a mix of people engaged in their own expressive cultural, environmental, and pyropolitical memory-making. Attention to these alternative techniques, I shall argue, calls for serious consideration of decolonization maneuvers within e-waste ethnography itself. On this note, this chapter explores the extent to which citizen[3] photography and similar participatory visual research efforts augment contemporary toxic studies in general and e-waste citizen science studies in particular. Attuned to the visual promises, politics, and possibilities of photography in toxic landscapes (Davies 2013, 2018), the chapter contends that engaging with visual citizen science techniques and like forms of participatory visualization and documentation can provide vital contextualization for debates grappling with the toxic injustices and environmental politics of e-waste labor. I explore how and why visual techniques in participatory action research matter in global environmental justice studies in general and postcolonial e-waste studies in Ghana in particular. As will be argued here, my emerging participatory e-waste visualization project, which rests at the intersection of citizen science and environmental studies, can push beyond the massive archive of contentious natural and humanist photography focused on Africa and the "prism of misery" (Kean 1998, 2) that too often typifies transatlantic and north-to-south visions of environmental destruction in Africa.

For most who have turned to Agbogbloshie as a site of e-waste research and visual art, there is an effort in all these projects to somehow extend an ethos of care for those living and working in this now notorious West African hot spot of "toxic colonialism" (Koné 2009).[4] But what seems less common in projects focused on Agbogbloshie are efforts to showcase how these workers are creative postcolonial agents actively documenting and

---

[2] For brief, yet concise, reviews of the emergence of and problems with the "ruin porn" concept, see Liboiron (2013, 2015).

[3] I follow the approach to "citizenship" taken up by Ellison, which suggests that citizenship is "a form of social and political practice born of the need to establish new solidarities across a range of putative 'communities' as a defense against social changes which continually threaten to frustrate such ambitions" (1997, 712).

[4] One of the first writings on e-waste dumping in Africa appeared in the late 1980s (see Brooke 1988), which emerged amid other works on "toxic terrorism" in Africa writ large (O'Keefe 1988).

communicating their own lived experience, pollution situation, and e-waste vitality. Kevin McElvaney's portraits come close, but that project is not an intentional photo-ethnographic project that aims to re-envision the narrative of Agbogbloshie from the standpoint and lived experience of e-waste workers themselves. In this chapter, I take issue with visual politics of representation in Agbogbloshie by engaging the following questions: What happens when e-waste workers are involved image-makers in efforts to represent Agbogbloshie? To what extent can this alternative visualization transform understandings of a place and space that has become a central node of global e-wasteland narratives? How does this twist to e-waste ethnography not only teach us in new ways but also provide an alternative educational foothold or platform to engage Ghana's e-waste truths?

Workers in this site, for example, have too often been understood as e-waste recycling laborers who foreground their experience of pollution and environmental health risk. But representations of Agbogbloshie as a site of e-waste toxicity and ruination are not the only stories being told. In light of this, the chapter navigates how e-waste "perceptual regimes" (Poole 1997) can be reconfigured and meshed with the overlapping projects of environmental justice and citizen science. It will be argued that e-waste workers and the images they produce constitute a process of perceptual inversion and representational plurality (Bleiker and Kay 2007). They teach us new ways of seeing and visioning e-waste contextualization and perhaps even the environmental justice challenges experienced in Agbogbloshie (Akese and Little 2018). My aim here, then, is to turn to a participatory visualization project in Agbogbloshie to stimulate critical discussion of the ways in which alternative e-waste visioning can transform how e-wasteland politics in Ghana are told, seen, and responsibly contextualized. Following Davies (2018, 549), I contend that "photography, despite its shortcomings, can breathe new life into the challenging task of making pollution knowable." Furthermore, it will be argued, it is important to foreground the power of photography even amidst contemporary social critiques of post-truth politics.[5] The truth is, "the power of the still image remains undeniable, even if some choose to ignore inconvenient truths" (Lyons 2017, 2). This reorientation of the power of visualization can provide a platform for "making new sense" (Hastrup

---

[5] One of the most powerful uses of photography-for-proof in recent times was on January 20, 2017, the nauseating day of Donald Trump's inauguration. Visual proof concluded that Barack Obama was the obvious victor of recent inaugural turnouts, despite the Trump administration's efforts to convince the public otherwise.

1995) of the complexities and confusion that shape contemporary and future representations of Agbogbloshie.

Refocusing on Worker Images

While acting as a primary environmental health response agency, PureEarth. org has also been an active player in circulating images that frame Agbogbloshie as a unique site of toxic pyrocumulous activity. In image after image, Agbogbloshie is seen as a burning place. NGO images and photojournalistic accounts help produce and sustain this dominant e-pyropolitical narrative. But despite the visual and narrative power of widely circulated e-waste "crisis" accounts, workers in Agbogbloshie, for example, describe and perceive this place as a "market" and place of work rather an e-waste landfill. More importantly, e-waste workers themselves engage in their own environmental risk visioning and visual documentation. They themselves capture and draw attention to images of toxic suffering in sequence with now classic images of ignited computer monitors, and they document radically different things and lived experiences. That difference makes a difference, especially for how one comes to understand, visualize, and represent toxic e-waste suffering and social life in the urban margins of what has become a narrative epicenter of e-ruination in Africa.

During ethnographic fieldwork in Agbogbloshie in the summer of 2016, I was talking to a group of workers about the risks they face when burning e-waste to extract copper. We sat under a makeshift shelter made from the roof of a junked *trotro*, the minibuses many Ghanaians use to travel from place to place. They began by talking about the smoke and how it impacts their breathing and sleeping, but they also started to show me their hands, feet, arms, and legs. They pointed to wounds, especially burns, on their bodies. Following this fieldwork experience, workers began using their cell phones to snap pictures of their wounded bodies. They would send me images of their cuts and burns on a weekly basis, and the more they shared these images with me, the more I started to realize that this practice of image-sharing was changing my own vision of Agbogbloshie and the workers making a living there. Their burnt and cut bodies became a medium of self-expression. Sharing pictures was a way to make sure I knew that bodily harm was still ongoing. It also became a visual platform from which to rethink the agency of bodies and subjectivities in Agbogbloshie.

When e-waste workers engage in this toxic risk documentation, they are actively contextualizing, communicating, and indexing the social and environmental health risks they face. One might call this a process of vital e-waste contextualization. Showing and marking vital signs of toxic risk contextualizes the experience of toxic burden, victimization, and marginalization within Ghana's e-waste recycling sector. But as visual anthropologists have pointed out, image-making is a political practice with various consequences. Reminiscent of the "crisis of representation" (Marcus and Fischer 1986) critique emerging in the 1980s in anthropology, multiple problems and politics of representation emerge when privileged outsiders (usually White and from the Global North) create representations of marginalized populations (usually dark and from the Global South) in order to expose inequalities and disparities in social, political, and economic systems. There is an overwhelming sense that image-making always risks being a practice of image-*taking*, a way of doing representational appropriation in a "contact zone" (Pratt 1992) where imperial and colonial power relations persist. All representational strategies, the general critique goes, directly or indirectly serve the interests of the image makers, no matter the shifts in representational ethos and edits to practices of objectification and dehumanization. In this way, upon forging and finding new ways to represent and visualize our toxic and polluted world (Davies 2018), we confront, no matter our representational strategies, the risks of "hazardous aesthetics" (Rosenfeld et al. 2018) and the enduring challenges of "visual interventions" (Pink 2007; Harper 2012; see also Harper 2002).

More recently, visual anthropologists contend that "relying on images of suffering bodies as a visual strategy of depicting injustice or inequality is at odds with making systemic social, economic, and political oppression visible . . . images of suffering bodies tend to naturalize connections between violence and already marginalized peoples. Furthermore, they do not ultimately work to make structural violence visible by (1) obscuring the mechanisms and perpetrators of violence, (2) not disrupting dominant conceptual frameworks, and (3) not leaving room for solutions" (Stone 2015, 179). Beyond simply a technology of documentation, taking photos can have intended and unintended self-serving consequences that can dehumanize e-waste laborers in Agbogbloshie, even for photographers emphasizing humanistic portraiture to help re-envision the e-waste narrative in the Global South. For example, during an interview focusing on his project *Agbogbloshie: Digital Wasteland*, the German photographer Kevin

McElvaney explains, "At first, Agbogbloshie caught my attention because it's really photogenic, but the environmental, socio-economic, political and ethical problems there forced me to see it with my own eyes," adding that "I don't like to judge things when I haven't seen it for real and everything I found about Agbogbloshie on the web seemed so unreal, but after I'd been there, I realized that it's even worse" (Donson 2014).

What I want to suggest here is that e-waste worker images, as a representational strategy for exploring environmental health risk, labor, and life in Agbogbloshie, provide new forms of everyday life documentation and cultural vitality that make Agbogbloshie a place that is *more than* e-waste recycling disaster and fiery crisis. What these workers face is what they digitally capture, and what they face can be many things; but it is rarely the burning pile of copper wires that constitutes the dominant optic of circulating narratives on Agbogbloshie. Each worker image I receive inspires me to ask what workers key into, what they account for, and why they share the images they share. Additionally, this grassroots e-waste visualization experience has inspired me to take more seriously the extent to which this digital documentation can provide a new platform for grassroots ground-truthing (or citizen science) and environmental justice approaches in postcolonial toxic landscapes, especially at a time when emergent forms of instrumentalization, computation, and digitization are shaping how we make sense of environment itself (Gabrys 2016). In this way, my interests in e-waste visualizations draw inspiration from recent discussion and theory in "digital anthropology," or doing ethnography in a contemporary technocapital world that "accounts for how the digital, methodological, practical and theoretical dimensions of ethnographic research are increasingly intertwined" (Pink et al. 2016, 1). Ethnography is now fully wired and technologized, a fact that certainly holds true for ethnographies of toxic e-wastelanding.

Peering through the dark and lethal smoke, it is critical to remember that Agbogbloshie is Ghana's most active scrap metal market. A regional and global node of supply-chain capitalism, it is where iron, steel, brass, copper, and aluminum are extracted and sold to supply and propel mega-development in India and China, especially development projects with a mega–electrical and electronic equipment footprint. (While all the burners tell me their copper and aluminum go to China and India, no research has actually explored where exactly these recycled metals from Agbogbloshie end up after they go to the port of Tema.) The social life of metal extracted and traded from Agbogbloshie needs to be actually traced and followed in order to really know where it ends up and what reuse economies these metal extractions support. What we do know is that wherever "development" is happening, copper is always an element of "development" infrastructure. The workers making up the informal labor force here come mostly from villages in northern Ghana, where economic hardships caused by colonialism have long been known to be a stimulus for southward labor migrations (Plange 1979). When these workers turn the camera on themselves, they purposively emphasize their copper wire collections. They take selfies in front of toxic flames. In one image the workers sent me, they are posing with prescorched copper wires; objects with market value are made to be the center of attention. As Yakubu (or Jacob), one of the longtime workers among this group of copper burners, told me during my visit in July 2017, "Copper is why we here. We be here for dat. All dis be copper. No work in north. This be work." Given this dominant metal in the scrapyard, it is easy to see why copper is foregrounded in this image. Copper is also a major reason why these workers migrate to Agbogbloshie from their villages in the north. When this image was taken, the price of copper was US$2.10 per pound, a commodity that is significantly more valuable than the ground nuts that many of these workers' families farm in Ghana's Northern Region. They also use images to communicate their humor and ability to goof off (see Figure 5.2).

For me, these e-waste worker images are not simply additives to my ethnographic approach but instead represent "an emerging platform for collecting, exploring, and expressing ethnographic materials" (Hsu 2014, 1). They help create critical dialogue and reflexive commentary on Agbogbloshie as a shared space and place with diverse active, and often contradictory, forms of representation and visualization. Many, if not most, ethnographies of pollution and environmental justice today involve what I would call a complex political ecology of data, what some in the digital humanities call our age of "augmented empiricism" (Hsu 2014) and what some Trump critics signal as an age of "environmental data justice" (Dillon et al. 2017). Of course, the turn to the visual and visual data is nothing new in the social sciences, nor is it new for discard studies focusing on the global e-waste trade and its local manifestation in spaces and places like Agbogbloshie. If the ethnographic use of these images risks reproducing colonialism and racist structures (Benjamin 2018; Comaroff and Comaroff 1992; Poole 1997), then what doe e-waste participatory photography actually do differently? What do these e-waste worker images still miss and overlook? What do these images produce and reproduce? What visual standpoint epistemologies and ontologies

Figure 5.2. Goofing off in Old Fadama. Photo by Abdrahaman Yakubu.

do they offer?[6] Perhaps more important, what politics of representation do or can these worker-generated images circumvent? How might visual decolonization become a technique of vital contextualization that can augment understandings of labor and social life in Agbogbloshie and other e-wastelands?

---

[6] See Haraway (1988) and Harding (1991) for foundational discussions of similar lines of feminist epistemic inquiry. Perhaps even more fitting here is a "feminist decolonial environmental justice" perspective (Murphy 2018, 101).

Figure 5.3. Scrap worker meeting and blessing in Agbogbloshie. Photo by Ibrahim Akarima.

Other worker photos help communicate and translate ritual blessings of the scrap metal market itself, a topic that no studies of Agbogbloshie to date have accounted for. Agbogbloshie workers have a primary meeting place for metal market meetings, negotiations, and blessings that they call "Gaza." As shown in Figure 5.3, one worker named Ibrahim took my camera and waited for a good time to take a shot during a meeting and blessing at Gaza. I asked Ibrahim what they were focusing their blessing on. What I learned was that this meeting had nothing to do with e-waste. It had nothing to do with praying for greater pollution control or federal waste management policy and action, but instead the gathering occurred to discuss and voice worker opinions, concerns, and ideas about the social, political, spiritual, and economic management of the scrap market. As Ibrahim told me, "The workers want the copper here. The iron and the aluminum here. We bless it so it come here." After taking a few more pictures of the meeting and blessing, Ibrahim repeated "They be blessing the copper. Blessing the market. Praying for *Naawuni* [God]."

They were not debating and discussing lead and cadmium exposures and various toxic body burdens of their labor, nor were they organizing a meeting to respond to NGO interventions to control pollution emitting from the burning of electronic discard in Agbogbloshie. Again, this "other" story matters. Even while critical relations between toxic bodies and environments (Roberts and Langston 2008) ought to remain a concern of future e-waste studies and environmental social science in Agbogbloshie, these relations must also consider the dominant role of Muslim culture, religiosity, and hereditary chieftaincy dynamics at play.[7] This is what sustains social life in the scrap market and what ultimately informs north–south labor migrations, movements, and experiences of people navigating e-waste risks. Inspired by Beck's (2006) idea that knowledge of risk and techniques of showing and visualizing risk are co-productive processes, we need to make on-the-ground facts visible. Our techniques of e-waste visualization need to come to terms with the actual suffering experienced by and among those making a living in Agbogbloshie but in a way that also tries to go "beyond the suffering subject" (Robbins 2013).

Still images of e-waste workers in Agbogbloshie do many things. They communicate postcolonial waste management and inequities in the global toxic waste trade but also friendship, tribal relations, and bodily distress. The latter topic has had a strong focus within photo-ethnographic work in medical anthropology. For example, medical anthropologist Paul Farmer has been criticized for his use of images of sick bodies to make visible what he calls "structural violence." He admits "the use of such images is problematic but sometimes necessary in order to stir privileged populations to do something about global systems of inequality" (Stone 2015, 180). The logic of this angle is that "the problem of making structural violence visible is that social, political, and economic structures that are to blame for the violence are very difficult to photograph because they are very difficult to see" (Stone 2015, 180). This is exactly why attention to e-waste optics matters. One could argue that citizen sensing and citizen visioning practices (like photography) don't necessarily "capture" experience (images are powerful but don't replace bare life experience itself), but they do "expose" or "share" lived *context* to the world of observers and those bearing witness. Furthermore, images can and often do generate felt empathy. When workers share photos, they are in some

---

[7] Attention to these factors is also highlighted by Stacey and Lund (2016) in their research on the dynamics of governance practices in the informal settlement of Old Fadama.

fast, digital way "sharing" their life experience with me. These images are shared by the so-called victims of e-waste toxic colonialism, but a problem emerges when images of suffering bodies are deployed "to illustrate injustice or structural violence locates all of the violence, the shame, and the danger of the violence in the suffering body of the victim rather than the assailant for the simple reason that the assailant is nowhere to be seen" (Stone 2015, 183). But what happens when postcolonial subjects are the image makers and agents of contextual revitalization? What happens when Agbogbloshie workers themselves visualize "landscapes of affect," when they themselves enable us to better understand "the simultaneous imagination and fabrication of inner selves, social bodies, and environmental milieu" (Moore, Pandian, and Kosek 2003, 31)? These are important questions for not only e-waste studies in the Global South but also environmental ethnographies of environmental justice, urban marginality, and discard writ large.

Drifting Toward a Decolonized E-Waste Ethnography

In this chapter, I have tried to show how in the process of making and sharing images, e-waste workers in Agbogbloshie are engaging in visualization practices that invert and confuse the dominant e-wasteland narrative in Ghana. As an ethnographer *being shown* what matters, and especially what matters beyond the "singular story" (Mkhwanazi 2016) of toxic digital destruction and extraction so common in representations of Agbogbloshie and the lives of those who work there, I have begun to find these images to be necessary tools for making sense of Agbogbloshie. These workers turn to images, including selfies, to share what matters to them. Accordingly,

> Rather than paralyze representational practices . . . visual depiction of structural violence need not settle for a qualified visual strategy heavily bolstered by written or spoken analysis. Reflexivity [in toxic studies] is a good strategy for many reasons, but it is not the only option. As the robust traditions of feminism and visual anthropology have argued, we should take the lead from the marginalized peoples who already work to make the abstract forces of [toxic] structural violence visible. (Stone 2015, 180)

How these workers show and share their experience of toxic suffering and violence is complicated by the fact that they don't necessarily turn the camera

toward the postcolonial state nor toward objects of toxic destruction to communicate their "environmental" experience. For example, the workers don't
send me images of urban authorities evicting workers attempting to dwell in
the Agbogbloshie scrapyard or shelters being demolished in Old Fadama, the
adjacent settlement, to mitigate the risks of annual flooding along the Korle
Lagoon. In short, the e-waste worker images validate an on the ground lived
experience that can be overlooked and lost amidst the prism of toxicity and
misery images shaping representations of Agbogbloshie.

Witnessing e-waste through participatory photography, as I describe in
more detail elsewhere (Little 2020), speaks to an ethos of "pluralist photography" attuned to power and methodological decolonization (Smith 2008;
Msila 2020). As Bleiker and Kay (2007, 158) put it, "Human relations cannot
exist outside power. But the nature of pluralist photography minimizes the
oppressive effects of these relations by consciously problematizing representation. The collaborative and dialogical nature of pluralist photography
can provide ways through which multiple perspectives may be seen and vali-
dated." In light of this, I want to highlight what I have found to be the benefits
and challenges of citizen visioning and collaborative photo-ethnography
in Agbogbloshie. To start, taking photographs is not a difficult thing to do.
There is a certain ease to image-making in our digital age. In this way, taking
pictures is an easy citizen science technique that even the poorest and most
marginalized of the Global South can engage. Worker-based photography
opens up e-waste ethnography to more grassroots and decolonized methodologies. On the other hand, the turn to grassroots citizen photography does
not escape criticism or necessarily lead to a more ethical ethnographic project. In fact, anthropologists have rightfully cautioned that visions of marginality and the practice of recycling the production of these representational
strategies and visions can sustain marginality itself (Ferguson 2006). Some
have even engaged this important issue more directly in Ghana by encour-
aging consideration of the always contentious nature of ongoing "Black
transatlantic visions" (Holsey 2013) and Black decolonial struggle (Mbembe
2021; Getachew 2019). These are research and representational concerns
that have informed my way of thinking about and doing participatory photography in Agbogbloshie. Working in collaboration with Agbogbloshie
workers to visualize Agbogbloshie was never my intended goal, but it became a focus once the workers themselves began to voluntarily snap shots
and send me images of their own making. In the process, I have experienced
a personal transformation in how to relate to these workers and their role in

my own ethnographic narrative. Their strategy is simple: they use images to help me find a way to understand and possibly make sense of Agbogbloshie. In the process, they have helped me understand my own self-critical stance on and understanding of the ethics of research and representation that emerge when navigating politics of pollution, participation, and the various environmental injustices of global electronic discard.

As I have noted, most depictions of Agbogbloshie's e-waste workers highlight toxic labor practices (e.g., burning copper wires) and do so to expose the contentious nature of e-waste dumping on the Global South. These e-waste representations, it turns out, have a charismatic quality and symbolic force that has consequences for contemporary e-waste theory and action (Lepawsky 2018; see also Liboiron 2016). In other words, "what makes e-waste charismatic is its capacity to act as an allegory of contemporary environmental crisis" (Lepawsky 2018, 6). The e-waste politics emerging in Agbogbloshie, then, are as much about environmental health crisis and toxic extraction as they are shaped by a complex and legitimate crisis of representation. What I am hoping to illustrate in my turn to workers' photographs of life and labor in Agbogbloshie is that these visualizations showcase the actual involvement of postcolonial agents in documenting their own lived experience.[8] They also provide an example for how e-waste workers visually intervene on their own toxic e-waste situation. This other technique of risk visualization can help advance critical and creative environmental health studies, synthesize the democratization of science and environmental justice advocacy efforts, and push the boundaries and intentions of ethnography in sites of toxic electronic discard more broadly. While grassroots citizen photography complements existing citizen science techniques, this form of visualization might also lead to more creative research and action partnerships and epistemic possibilities within a twenty-first-century mixed-media environment that "has made the landscape a stage for human drama" (Adelman 2020). It can also help us reimagine the actual performance of the "shifting ecology of citizenship" (Comaroff and Comaroff 2016, 29), a complex lived process involving fluctuating experiences of precarity, inequality, and security, especially among the urban poor (Das and Randeria 2015). While liberal theory might assume that citizens engage passively with scientists to document everyday life in Agbogbloshie,

---

[8] Emerging scholarship on toxicological science and contamination politics in Senegal similarly highlights this need for greater attention to postcoloniality (Tousignant 2018).

this does not mean "scientific engagement makes citizens" (Leach, Scoones, and Wynne 2005, 13). Instead, if anything, Ghana's e-waste workers practice a form of urban citizenship recognition that springs from a perspective of the citizen as "a more autonomous creator and bearer of knowledge located in particular practices, subjectivities and identities" (Leach, Scoones, and Wynne 2005, 12).

But my intentions in this ongoing ethnographic project are multifaceted and challenged by my own process of self-realization and growing sensitivity toward decolonization. Without knowing or having ready-made answers and solutions, I still find it critical to continue trying to carefully learn how to make e-waste ethnography and e-waste worker advocacy work in postcolonial Ghana more ethical. What I do know is that attending to the "primacy of the ethical" (Scheper-Hughes 1995) has meant for me a slow meditation on what e-waste ethnography in Ghana is or can be, not necessarily what it should be or do. Engaging in unanticipated participatory photography not only helps capture the visual injustices of global e-waste but also reinforced, for me, that ethics of research and representation are in fact refined through ethnographic practice.

In light of this reflection, a visual e-waste ethnographic approach also fits within certain parameters of efforts to "decolonize methodologies" (Smith 2008), not simply efforts to insert research produced by e-waste workers into my own research fold. Following the insight of the Maori indigenous scholar and activist Linda Tuhiwai Smith, we are reminded that "Decolonization is a process which engages with imperialism and colonialism at multiple levels. For researchers, one of those levels is concerned with having a more critical understanding of the underlying assumptions, motivations and values which inform research practices" (Smith 2008, 20). Additionally, one can't honestly discuss decolonization without also engaging the orbiting force of racial politics that inform (or should inform) ethnographic reflection. As anthropologist Jemima Pierre (2013, 551) rightfully notes,

How do we, in fact, analyze the ways that Africans (in former non-settler colonies) continue to grapple with issues of white power and privilege? Postcolonial Africa's engagement with whiteness and discourses of race, racial difference, and privilege occurs within a broader set of processes whereby local relationships continue to be structured by the current global configurations of identity, economics, and politics. These processes demand radical racial analysis.

To date, e-waste ethnography in Africa in general, and in Agbogbloshie in particular, has developed few direct ties to or dialogue with such critical race studies and forms of analysis, nor have e-waste studies in the Global South dealt directly with emerging interests in alternative Global South epistemologies.

In May 2018, I attended a joint conference of the American Anthropological Association and the African Studies Association in Johannesburg, South Africa. At the opening session of the conference, participants were challenged to use the three-day event to ponder a critical question: Who could come up with the best concept for explaining the current state of scholarship *in* and *of* Africa today? The chosen concept was announced at the conference's closing session. "Epistemic decolonization" was the winner. It was suggested that the concept allowed for a more radical epistemological practice that seeks to invert and reroute the process by which knowledge is produced, including *who* in fact produces knowledge and *where* such knowledge is produced from and originates.[9] In this framework, Africans need to be recentered in projects to make Africa knowable. This was also the subject of a Ghana Broadcasting Corporation program I watched one evening during fieldwork in Ghana in July 2017 called *Science in Africa*. The reporter on the program noted that "The problem is that knowledge and technologies to solve Africa's problems are imported from outside. We need to have Africans and African minds solving African problems. This will come down to sustainable interest in science as not a problem solver, but as a knowledge source that can be used to push Ghana forward" (Ghana Broadcasting Corporation, 2017). In the case of Agbogbloshie and e-waste, knowledge produced by e-waste workers is a critical epistemic infrastructure, not some "local knowledge" or additive to my own ethnographic research project and knowledge production practice. Echoing recent anthropological inquiries into infrastructure, we might seriously consider the following questions to rethink and decolonize e-waste studies and engagements: "What kinds of infrastructure—epistemic, energetic, political—might we contemplate from the everyday ruins and rubble wrought by infrastructure today? How might we reimagine their forms and

---

[9] As Santos (2014) might suggest, epistemic decolonization is basically a fight against "epistemicide." For another thread in this debate, see Comaroff and Comaroff (2012), as well as Jean Comaroff's comments delivered at the "Decoloniality and Southern Epistemologies" webinar sponsored by the African Studies Global Virtual Forum on November 13, 2020 (https://www.youtube.com/watch?v=5OGb-DRBqqk&t=953s).

potentialities anew in times where the end of life itself has been rendered thinkable?" (Appel, Anand, and Gupta 2018, 30). Places like Agbogbloshie exist as not only sites of technocapital ruination where juggernaut technological obsolescence is materialized but also sites of productive e-waste infrastructure and circular economic development. In this way, more than a toxic "mitigation landscape" (Little 2014), Agbogbloshie is an infrastructural landscape where a metal recycling market serves as a primary source of income, a place where material continues to come and go and where burning labor breaks down discard. For workers it is a constant, yet precarious, infrastructure of cash but also a place with its own knowledge economy. Agbogbloshie is a space and situation they know, deep down, more than anyone else. Because of this, I often find myself struggling to really know what to say and know about Agbogbloshie. No matter the exciting epistemic foothold that a more just and decolonized e-waste anthropology might seem to offer, it seems that we can't really talk about epistemic decolonization or even visual decolonization as methods to advance e-waste ethnography in the Global South without also taking a harder look at who actually benefits from this epistemic deconstruction effort. While deconstruction is a necessary first step in any decolonizing effort (Smith 2008), it too often falls short of actually having any positive outcome for communities that have suffered from colonization and imperialism.

I would be wrong to suggest that my effort to re-envision e-waste ethnography in light of decolonization tactics will lead to a better life for Ibrahim and his fellow workers. But I also have found that both being sensitive to and even just trying to decolonize methodologies deployed in e-waste ethnography in polluted places like Agbogbloshie is a step in the right direction, even if it is incomplete and filled with representational politics. Ultimately, the front-and-center challenge is to sustain "the importance of looking at structural, slow, and epistemic violence in unison" (Davies 2019). As more critical ethnographies of e-waste and discard extraction emerge in the years to come, we might expect greater clarification, realization, and understanding of global neo-/post-/colonial processes and their significance to the practice of future global e-waste ethnography itself. What remains certain is that Agbogbloshie is a precarious e-wasteland of ongoing neoliberal experimentation, where toxic bodies, commodity metals, and contaminated ecologies are the target of desperate, yet optimistic, projects, interventions, and infrastructures that, for now, do little to actually protect the livelihoods of the Dagomba workers who make up Ghana's urban e-waste labor core.

6

Looming Uncertainties of Neoliberal Techno-optimism

All around the body reigns an atmosphere of certain uncertainty.
—Frantz Fanon (1952, 90)

If they bring the wire stripping machines, the burning will stop here.
—Chair, Greater Accra Scrap Dealers Association

Agbogbloshie, as an epicenter of global e-wasteland narratives, is part of a globalization process that flourishes in fragmented and messy ways. Here we find not a ready-made e-waste problem but instead fragments of a problem that require reckoning with how and in what ways these fragments make sense and to whom. Following Tsing (2015, 271), the e-waste problem in Ghana "is not delivered whole and round like a pizza, to be munched and dismantled by the hungry margins. Global connections are made in fragments—although some fragments are more powerful than others." In light of this phrasing, we can think of e-waste friction in Agbogbloshie as a fragment of global e-waste management and global environmental health science, a two-pronged fragment of "global" connection and intervention in a place that has drawn the attention of intermediary domestic and international nongovernmental organizations (NGOs), "world-thinking" (Mbembe 2017, 179) agencies born from and empowered by post–World War II global reordering.[1] One NGO project sought to even "eliminate" the burning of electronics, an interventionist logic which has pyropolitical

---

[1] Following the development of the United Nations in 1949, the term "nongovernmental organization" was originally used to describe "agencies that would remain at a distance from governments, acting as their conscience and offering a moral critique of states. Since then, however, this mandate has expanded as NGOs have exploded in number, size, and scope" (Leve and Karim 2001, 53). These agency formations have, of course, been a centerpiece of numerous social science critiques (see Townsend, Porter, and Mawdsley 2004; Gupta and Sharma 2006; Schuller 2009).

Burning Matters. Peter C. Little, Oxford University Press. © Oxford University Press 2022.
DOI: 10.1093/oso/9780190934545.003.0007

implications as these interventions on e-waste burning often include environmental health logics and goals that can, and often do, clash with the basic socioeconomic needs of Ghana's working poor. This is a primary reason why Agbogbloshie and Old Fadama have seen the emergence of numerous civil society organizations. As the following list of organizations indicates, toxic e-waste is hardly the sole focus of local organizing and community advocacy interests. Groups working in the area include the Slum Union of Ghana, People's Dialogue, Agbogbloshie Scrap Dealers Association, Old Fadama Community Development Association, Ghana Federation of the Urban Poor, Informal Hawkers and Vendors Association of Ghana, Green Advocacy Ghana, You Caring, Good Electronics, Closing the Loop, Agbogbloshie Makerspace Platform, Yam Sellers Group, Tomato Sellers Association, Truck Owners/Operators Association, and Susu Collectors' Union. The very fact that this local organizational diversity and focus exists suggests that there is a problem or crack in the dominant e-waste spectacle. Given this plurality of extra e-waste problems, we need to attend to the ways in which e-waste intervention logics and infrastructures, and the techno-optimism used to justify their deployment, overlook critical "fragments" that in fact matter.

In this chapter, I argue for an alternative approach to e-waste intervention in postcolonial Ghana that, in fact, inverts the focus and emplacement of intervention itself. In particular, I ask, what if intervention in Agbogbloshie takes place elsewhere and not in the toxic hot zone of e-waste cable burning, as it has been? In what ways would this approach help decontaminate NGO visions and practices and perhaps even help demystify current e-waste representations? Surely, the scrapyard is where toxic emissions occur, but if environmental health intervention and its exposure science praxis is meant to reduce actual exposures and mitigate bodily accumulations of toxics, finding ways for these migrant laborers to make a living in their villages in the north is one of the most effective ways of actually reducing unnecessary chemical exposures for those laboring in Agbogbloshie. In this way, this chapter suggests a turn toward a more radical and just environmental health intervention approach informed by rural–urban migration logics and the challenges and necessities of rural livelihood sustainability. I shed light on the ways in which recent discussions of epistemic decolonization and "just transitions" (White 2020) can help reconceptualize e-waste politics and solutions in Agbogbloshie and the transgressive nature of ongoing NGO and state interventions there.

Efforts to decontaminate Agbogbloshie and make e-waste recycling labor more "healthy" and e-waste recycling infrastructures more "green" became most explicit in 2014, when Agbogbloshie saw the development of an experimental e-waste recycling facility that hoped to modernize and formalize an expanding informal scrap metal trades sector. The goal of the facility, which was later named the Agbogbloshie Recycling Center (ARC), was to make e-waste recycling work safer and more environmentally friendly. In partnership with the European Commission, the United Nations Industrial Development Organization, the Global Alliance for Health and Pollution, Ghana's Environmental Protection Agency (EPA), Ghana Health Services, Green Advocacy Ghana, and the Greater Accra Scrap Dealers Association (GASDA), the project has been mostly designed and financed by Pure Earth (formerly known as the Blacksmith Institute), a solutions-based environmental NGO that aims to "identify and clean up the poorest communities throughout the developing world where high concentrations of toxins have devastating health effects." The primary mission of the e-waste recycling center was straightforward: "Eliminate Burning at Agbogbloshie." With support from a variety of government and nongovernment agencies, the new recycling facility has a clear risk reduction goal. But the ARC facility also aimed to automate e-waste labor, a move reflective of Ghana's so-called second age of optimism (Droney 2014), a time of renewed interest in building infrastructures of science and technology in Ghana. But all these e-waste innovation and upgrade practices in Agbogbloshie expose the broader "problem of capacity" (Tousignant 2018) in a precarious postcolonial setting of uncertain infrastructural and techno-developmental promise.[2] The truth is that e-waste interventions in Agbogbloshie are scarce, stagnant, and uncertain, even though these interventions and science projects involve a certain ethos of techno-optimism, even a techno-future ethos that many Ghanaians themselves share. As Droney (2015, 224) notes, "Places like Silicon Valley are characterized by boundless, blinding optimism that sometimes seems like an unshakable faith in the inevitable (and profitable) unfolding of technological progress as an inherent good. But this is not limited to California; places like Ghana are a part of this same culture of techno-optimism." As argued here, the ARC is best understood as an intervention infrastructure "around which political debates coalesce . . . where the present state and future possibilities

---

[2] Tracking toxicologists and environmental health scientists in postcolonial Senegal, Tousignant (2018) unearths a story of sociotechnical science and intervention in conditions of scarcity, stagnation, and loss.

of government and society are held up for public assessment" (Larkin 2018, 177). In the case of the ARC, e-waste techno-optimism is also interrupted by workers who are rightfully skeptical of these "innovations," especially since automation mostly displaces human labor rather than integrates human labor (Benanav 2020).

Many questions inform my approach to e-waste intervention uncertainty and techno-optimism in Agbogbloshie, but the one that steers the focus of this chapter is, what happens when the techno-optimism driving risk mitigation interventions meets the practical challenges and social suffering of e-waste recycling? Guided by this question, this chapter has two primary goals. On the one hand, it explores ethnographic interview data that exposes how NGOs and managers in Agbogbloshie understand sociotechnical renderings of risk, e-waste, and the enduring and interrupted forms of pollution control and risk mitigation emerging in Agbogbloshie. Additionally, the chapter aims to reflect on e-waste ethnography to generate critical discussion of techno-optimism in a sociotechnical environment of significant capacity-building struggle and precarious neoliberal waste management. Finally, my goal is to illuminate the possible efficacy and struggle of e-waste interventions without losing sight of power relations shaping the angles and outcomes of interventions in postcolonial Ghana that usually recycle and perpetuate green developmentalist agendas, projects, and discourses of hope centered on cleaning up Ghana's dirty e-waste mess.

Unraveling Green Intervention

Agbogbloshie exhibits all the features of stagnant NGO intervention and the "(non)functionality of materials," as Tousignant (2018) would put it.[3] Stagnant e-waste interventions, in this sense, are conditioned by the entanglement of temporal, social, and economic realities that impede the very performativity and ebb and flow of NGO-driven techno-optimism. My goal here is not to drag yet another NGO intervention in Africa through the mud but instead to expose the complex muddle one international environmental NGO operates in and how this muddle matters to understanding the politics of techno-optimistic interventions informed by economic development,

---

[3] For excellent reviews and critiques of enduring NGO power in Africa, see Hearn (1985, 2002) and Stewart (1997).

alternative labor arrangements, and global environmental health science. There is no shortage of social science critiques of NGOs and pollution management politics in the Global South, but the case of techno-optimism across Africa has a particular history that spills over into current discussions of e-waste pollution control and global environmental health risk management in Ghana. What is important to recognize, as Hecht (2018) points out, is that the study of waste and waste management is an emerging and captivating domain of scholarship precisely because waste and waste management inspire new perspectives on social, political, and cultural relations, scenarios, and demands (Hawkins 2006; Gille 2007; Chalfin 2014; Reno 2015; Liboiron 2016; Harvey 2017). The ongoing e-waste interventions in Agbogbloshie have generated new relations between urban scrap metal workers, international NGOs, and state health and environmental agencies. Among recent techno-optimistic projects emerging in Agbogbloshie, there is a certain neoliberalization of e-waste intervention, a specter of renewing a metals recycling economy in a way that will be globally competitive and create new, formal job opportunities. Yet, all of these intervention projects are emerging at a time when the promises and pitfalls of capacity and modernization are on full display and sometimes, unfortunately, in near or full decay. Interventions on e-waste can, as this chapter shows, generate further uncertainty, especially uncertainty about real options and solutions to manage urban e-waste within a global system of "capitalist ruin" (Tsing 2015).

As argued throughout this book, I consider Agbogbloshie a complex e-wasteland and cultural space of not only toxic e-waste recycling and commodity chain extraction but also the e-burning that made it a place of e-wastelanding concern at all. It was, after all, the multiple environmental health risks of e-burning that became the central focus of mitigation solutions, infrastructures, and environmental politics in Agbogbloshie (Little 2016). In this way, Agbogbloshie presents a case for a certain kind of techno-optimism, a desperate techno-optimism taking shape in a postcolonial environment of extreme contamination and socioeconomic precarity. The concept of desperate techno-optimism is leveraged here to rethink e-waste interventions in a place and space dominated by narratives of e-waste dumping, desperation, and toxic misery. Treating e-waste management interventions in Ghana as matters of "sociotechnical imagination" (Jasanoff and Kim 2015) helps connect the micropolitics of e-waste to broader political economic and environmental trends. Ultimately, making these connections is one way to combat the all-too-common pattern of reducing Agbogbloshie to an unfathomable digital dump and of treating Agbogbloshie itself as a sort of runaway subject of modern discard, or perhaps, metaphorically, even a blazing "hyperobject" (Morton 2013) of toxic struggle that can feel unreal because of the very overwhelming nature of its environmental intensity.[4] In this way, the disaster imaginaries that have dominated renderings of Agbogbloshie have overlooked the consequences of sociotechnical "intervention" imaginations. I would suggest that sociotechnical imaginaries, in this sense, are intervention imaginaries that can expose the politics of optimistic decontamination efforts and risk mitigation technologies.

What most government and NGO proponents of e-waste intervention face in Ghana is a recognizable "future imperfect" (Jasanoff 2015), whereby the desire to do quality and sustainable intervention is met with both the capacity for imagination and the interruption of imagination. Based on my experience in the Agbogbloshie scrap metal market, there is an overwhelming sense that e-waste intervention projects will either turn to scrap soon or be scrap from the start. Interventions quickly become examples of resource depletion, infrastructural breakdown, and corrupt financing. Some projects even become "residue" in the very toxic environment they seek to mitigate (Boudia et al. 2018), while others implicitly recycle "sociotechnical imaginaries" (Jasanoff and Kim 2015), forms of imagination and materialization haunted by stagnation, dissipating funds, and operational failure. A recognized challenge is to rethink the problem and offer an analytical perspective that encourages us to think differently about e-waste risk mitigation and techno-optimism in a postcolonial Ghana confronting neoliberal problem-solving discourses and urban planning strategies, such as Ghana's National Urban Plan, which is squarely focused on "attracting and harnessing foreign capital support and inward investment, and, at the same time, [recognizing] the role of the informal economy in terms of businesses, markets, and settlements" (Oteng-Ababio and Grant 2019, 5).

The ARC, Ghana's first "state-of-the-art" e-waste recycling facility, promised to provide a solution to the toxic overload experienced by Agbogbloshie's laborers and those living in the surrounding area. Even though the campaign to "eliminate burning at Agbogbloshie" was presented as a multiagency and organizational partnership, the intervention approach was largely designed by members of Green Advocacy Ghana, also known as GreenAd. As the

---

[4] As philosopher Timothy Morton puts it, "The feeling of being inside a hyperobject contains a necessary element of *unreality*—yet this is a symptom of its reality!" (2013, 146, emphasis in original).

leading environmental organization in Ghana to directly address e-waste issues in Agbogbloshie, GreenAd "is an organization that aims at upholding and enhancing the sustainability and integrity of Ghana's environment" (Green Advocacy Ghana, 2021). Led by some of Ghana's most influential and talented environmental scientists, GreenAd strives to reach its organizational goals through "Partnership in research and data collation on the state of the environment; Development and the availability of environmental databases; Capacity building and dissemination of environmental information; Attitudinal nurture to reflect positive environmental understanding; Application of environmental knowledge to engender sustainable lifestyles and communities; and Promotion of and motivation for enforcement of laws and regulations" (Green Advocacy Ghana, 2021).

Shortly after I arrived in Ghana in July 2017, I arranged a meeting with members of GreenAd. We met at their air-conditioned office at Sakumono Estates near Tema, the industrial port on the eastern edge of Greater Accra where most recovered metal scraps in Ghana end up for processing and export. When I arrived, I was kindly greeted by one of the co-directors and handed a packet of information about GreenAd. While I waited for other officers of GreenAd to arrive in the meeting room, I sifted through the information provided and was drawn to a newsletter called *Green Alert*, a youth-focused environmental newsletter "intended to inform, inspire and involve young people in the campaign to reverse the many environmental ills which afflict our nation." Page two of the newsletter really caught my attention, a section titled "Environmental Degradation & Poverty: Two Sides of the Same Coin." The short article began with a tale that young Ghanaians had likely heard before. It is the tale of four blind men being asked to identify an object in front of them. Each man investigated one part of the object, so none of the men identified the object, which turned out to be an elephant. In other words, they missed out on "the whole." The article went on to explain how this metaphor is employed to reckon with the tight relationship between poverty and environmental destruction in Ghana, ending with "Efforts at fighting environmental degradation and poverty must never be separated. They are two sides of the same coin. It is therefore not a mere coincidence that environmentally marginal and degraded areas tend to have the highest incidence of poverty. In dealing with environmental degradation therefore, we would be getting rid of the underlying causes of poverty, for they are two sides of the same coin" (*Green Alert* 2010). Just like the blind men who don't identify the "whole" elephant, I want to single out, at least for the moment, the intervention actors in Agbogbloshie that draw attention to the problem of environmental toxicity since this has been a primary point of interest for GreenAd more recently.

GreenAd, like many civil society organizations and NGOs these days, has a Facebook page. Perusing the page in January 2018, I came across the following post:

> As the stream of E-Waste into Agbogbloshie continues to grow, the scale at which these practices are being used is expanding. Harmful practices include the burning of plastics, foam, and rubber—emitting incredibly dangerous pollution into the air, soil, and water. Laws against crude recycling methods used in Agbogbloshie have not been strictly enforced, allowing soil, air, and water pollution to reach alarming levels. Our hope is to formulate a permanent solution towards the wellbeing of the surrounding population and environment. (Green Advocacy Ghana, 2018)

One effort to generate a more "permanent," or at least long-term, solution was the development of the ARC, which culminated with a partnership between GreenAd and Pure Earth, an international "solutions-based" environmental NGO that since 2007 has been developing a list of the world's most polluted sites.[5] Yaw Amoyaw-Osei, who is the executive director of GreenAd, is also Pure Earth's expert in Ghana and has worked for many years with Ghana's EPA conducting research on e-waste. I have never been able to meet Yaw, but what I do know based on ethnographic research is that despite the combined efforts of GreenAd and Pure Earth, the majority of e-waste workers in Agbogbloshie see little help from these e-waste management "solutions." As Oteng-Ababio and Grant (2019, 6) suggest, most of these solutions and management practices end up being "both unpredictable and unsatisfactory for all," a situation that repeats the common cycle of uncertainty and "crisis" that leads to increased experimental intervention and NGOization across Africa. Moreover, as James Ferguson (2005) might suggest, another problem with this NGOization might be rooted in the fact that the actual "Earth"-scale focus of Pure Earth serves a certain neocolonial function: "The implication usually drawn from such invocations of the environment [and the e-waste

---

[5] Reports by the Blacksmith Institute/Pure Earth have been the subject of local skepticism and critique in some regions in the Global South (e.g., Peru), but efforts in Uruguay have resulted in more optimistic outcomes, like community-based lead poisoning research and surveillance (see Renfrew 2018, 217–218).

crisis] as a global issue is that the 'national level' is inadequate for environmental [and e-waste] regulation and protection, since environmental [and e-waste] crises do not respect national borders" (Ferguson 2006, 42).

"No Help" Narratives Amidst Intervention Celebration

As noted in Chapter 4, Pure Earth's health promotion campaign, while important as a strategy to raise environmental health awareness, exposes a "blame the victim" strategy with biopolitical consequences. The campaign aimed to offer easy-to-understand solutions to negative practices of informal e-waster labor, practices that have well-documented health impacts. E-waste workers were told outright that *Kona ne bad* ("Burning is bad") and that they should wear personal protective equipment and care about their health. But as an intervention combining self-surveillance and good health practices centered on individualized choice and "best practice" behavior, Pure Earth's approach to formalizing e-waste recycling in Agbogbloshie still leaves a burning question unanswered: "How can we ensure a 'middle ground' for continued access to livelihood without compromising environment and health?" (Amankwaa 2014, 2). Despite the efforts, this "middle ground" isn't met and can even lead to further marginalization of Ghana's e-waste laborers. This increased marginalization has many sources, but some highlights deserve attention. First, the workers who are targeted to benefit from this intervention don't have direct access to the facility as the facility is gated and under the control of only certain members of GASDA. Second, the vast majority of copper and aluminum burners aren't trained to use the granulators for stripping the wires. Again, a select few of GASDA's members have been trained properly. Third, even if a copper or aluminum burner had access to the facility and knew how to operate the granulators, this is a very slow process for generating raw copper wires. Therefore, to use the e-waste recycling facility would severely slow down the copper extraction process. It became obvious during my repeated trips to Agbogbloshie that the ARC facility struggled to sustain operations. Beyond the burden of basic utility costs, like purchasing electricity to run the granulators, electricity is not always reliable, despite Ghana's mega-investments in hydropower.[6] Also, as highlighted in a recent training manual

---

[6] On several occasions, I have been told that the sporadic availability of electricity and the daily planned outages throughout Accra fuel local criticism of Ghana's hydropower investments.

produced in partnership with Ghana's Ministry of Environment, Science, Technology and Innovation, "Machines for shredding of cables have been provided at Agbogbloshie by other institutions but workers are not using them due to maintenance reasons and time efficiency" (Deutsche Gesellshaft fur Internationale Zusammenarbeit 2019, 41).

Finally, the facility primarily processes waste electrical equipment in the form of large-diameter cables sourced directly from Ghana's electrical utility company, the Electricity Company of Ghana, Ltd.[7] This situation leads to machines sitting idle since most of the granulators cannot feed small-diameter cables, which are what copper wire burners work with most. All of this is to say that this "modern" e-waste intervention effort is met with confusion and frustration, despite the initial thrust of NGO optimism. Pure Earth's e-waste facility project operates under the "progressive" idea that it can be a win–win for both workers and the environment, but many workers I have interviewed are skeptical of the efficacy and sustainability of the project. No matter how formalized the e-waste recycling sector gets, I am told by one worker, "this facility not good for me or many of us. It be good for someone, but everyone here know that the government want us out. They evict our people [in Old Fadama] who work here. But we do the e-waste recycling. And you see, nobody care for us here." This was a common response during ethnographic interviews among Agbogbloshie workers. Further examples of what I am calling "no help narratives" include the following:

"No one here care. No money. No chop [food] money." (Interview, July 10, 2017)

"Where there be help? Where? Everyone here is hustling. Everyone trying to make chop money." (Interview, August 3, 2016)

"They will burn because they have nothing here. Nobody has the money to stop the burning. Blacksmith [now Pure Earth] came. We need the help. But what can we do? How can they help us here?" (Interview, July 16, 2017)

Those working on the margins of this e-waste recycling facility are rightfully conflicted, confused, and confident that this intervention won't directly benefit their livelihoods. This skepticism has even resulted in creative

---

[7] The utility company was founded in 1967, with the tagline "The Name Behind Electricity in Ghana."

Figure 6.1. Burnt information flier. Photo by author.

pyropolitical resistance. For example, one morning during a stroll through Agbogbloshie I noticed a burnt Pure Earth information flier taped to the wall of the ARC (see Figure 6.1). I never found out who did it, but whether it was an act of symbolic justice or just playful NGO vandalism, it does expose another thread in the pyropolitical ecology of e-waste intervention. If e-pyropolitics were an image, this would be it.

While there is no shortage of NGO critique to go around, especially NGO involvement in neoliberal Africa writ large (Ferguson 2006), Agbogbloshie exhibits a unique case where African bodies themselves are seen as the ideal target of environmental health intervention (Butchart 1998; Little 2019). Despite the importance of pollution mitigation and decontamination efforts, my ethnographic research points to an equally pressing concern among workers, which is their ability to work and sustain an economic livelihood that can support their families. E-waste workers know their bodies are on the toxic front lines. They occasionally weigh the options as best they can, but many explain that no matter how hard they try, they can't escape the poverty and marginalization they confront. "Nothing for me," Awata tells me. "Nothing. That be why we burn here. That be why I go for copper scrap."

When Pure Earth first opened the cable stripping facility in Agbogbloshie in October 2014, there was a grand opening celebration. In the words of Pure Earth,

> A celebration marked the opening of the new e-waste recycling facility in Agbogbloshie, Accra, Ghana. Dignitaries, dancers, performing school children and a balloon archway stood out in sharp contrast in this sooty, trashstrewn landscape. This collaborative effort to transform Agbogbloshie from one of West Africa's largest e-waste dumps into a model recycling center was launched on October 9, 2014. . . . The overflowing crowd at the opening reflects the strong community support for the project. (Pure Earth 2014)

Local dignitaries and government officials attended the event, including the deputy minister of the environment, the director of the EPA, and representatives from the World Bank, the United Nations Development Programme, Ghana Health Services, the National Youth Authority, the Ministry of Energy, the Ministry of Trade and Industry, the Electricity Company of Ghana, and many others. According to Sam, a scrapyard manager I interviewed in the summer of 2015, the event was well received; but its impact was uncertain: "Blacksmith [now Pure Earth] was here and they had the big program [opening ceremony]. Many many people came. We felt it was good to have new machines here. Because the burning is too much now. Too much smoke and sickness here. Blacksmiths was here to help. We don't know what will happen now."

After initial hesitation, the community of workers and especially GASDA seemed to be optimistic about this partnership, even if they garnered skepticism from previous attempts by the Ghana EPA to crack down on pollution emitting from the scrap metal market. When I returned for fieldwork in 2017, the Pure Earth facility was clearly on the wane. The building that housed the granulators had been repurposed, and there was no signage indicating "hours of operation" or available services. A pile of shredded cables stood 8 feet high in front of the entrance, and there was no sign of recycling customers. What I confronted was yet another example of the infrastructural decay and project debris of NGO boom-and-bust (see Figure 6.2). So, while Ghana's own environmental health experts (Asante et al. 2016) contended that "it is expected that the e-waste workers will embrace the use of these machines" and that use of the granulators "will go a long way to help prevent the burning of e-waste and will be replicated in other smaller e-waste centers in the country," what remains of the ARC intervention is rust and rubble. The exhausted ARC facility symbolizes the overwhelming nature of the scale and scope of scrap metal being processed and flowing in and out of Agbogbloshie. Production and reproduction, innovation and destruction, intervention and mitigation; these are the interconnected and inward-collapsing processes of enduring technocapital ruination. Together these processes illustrate the slow and fast violence of the matter (Nixon 2011).

Figure 6.2. The exhausted ARC facility. Photo by author.

New Infrastructural Promises and Uncertainties

There is little debate that the Agbogbloshie scrap metal market plays a central role in Ghana's urban industrial activity. Much like other regions of Greater Accra, this industrial region of western Accra has since the early 1980s become an "active frontier" of neoliberalism (Chalfin 2010).[8] It is a busy political economic space that inflects global trends, primarily "specific macroeconomic and macropolitical interventions geared to the promotion of private enterprise and the market logics strongly associated with economic globalization and the spread of multinational capital" (Chalfin 2010, 4). This economic globalization has created conditions for climate change friction. Much like neighboring countries in equatorial Africa that struggle with urban waste excess and management, black carbon– and greenhouse gas–emitting sites like Agbogbloshie have quickly become a target of climate change blame. Beyond the environmental health challenges that first attracted international NGOs to Agbogbloshie, Ghana's climate change challenges also helped turn a critical eye on the pollution emitting from Agbogbloshie. As noted in the Introduction, the Ghanaian state has responded to this challenge through active participation in international climate change negotiations and domestic planning through sustainable growth where possible. It remains uncertain whether the $22.6 billion needed to carry out its own "sustainability" pledge, a nationally determined contribution Ghana made as part of its obligations under the Paris Agreement, will ever come to fruition.[9]

New forms of techno-scientific expertise were called on to justify the development of the promised e-waste facility. To provide national, regional, and international stakeholders with "a summary of expert opinion on the most pressing problems arising from e-waste activities at Agbogbloshie, as well as suggested solutions to address these problems" (Cazabon et al. 2017), in 2015 research scientists conducted structured interviews in Agbogbloshie. They used what is known as a "logical framework approach," a scoping exercise used to gauge problems and benefits of e-waste recycling and identify

[8] Following Harrison (2011, 449), I agree that studying neoliberalization in Ghana requires site-specific knowledge before making neoliberal claims: "Globalized and bombastic as the neoliberal project might appear, its ability to generate social conformity in the diversity of states within Africa (let alone elsewhere) is not self-evident; rather it requires that we research the dynamics of changing practices of governance in specific sites before we generate broader commentary on neoliberalism's prospects."

[9] Ghana's sustainability goals align with the UN's Sustainable Development Goals, which went into effect in January 2016, and aim "to foster economic growth, ensure social inclusion and protect the environment" (United Nations Communications Group 2017, 3).

response options. As noted in Chapter 4, according to researchers working on this project, the primary goal was to prioritize potential interventions that would address identified problems at Agbogbloshie. These researchers concluded that in addition to there being a clear need for further research on the health effects of "primitive" e-waste recycling, e-waste workers should be given "appropriate personal protective equipment"; that the Ministry of the Environment, Science, Technology and Innovation revisit Ghana's Hazardous Waste Bill; that workers themselves be "involved in the planning process of interventions" and "be kept informed of any results"; and finally, that there be increased education and "sensitization" on hazards related to e-waste for both workers and the general public (Cazabon et al. 2017, 980).

A second round of techno-optimism is coming from the state's promise to develop a "state-of-the-art" recycling facility—one much larger and more heavily financed than the ARC—to put an end to toxic e-waste burning in Agbogbloshie and to reform Ghana's e-waste recycling sector in a way that positions them better within the global e-scrap circular economy (see Figure 6.3).

Figure 6.3. New infrastructural optimism. Photo by author.

Funded by SGS Limited, a French-based global certification company, and managed by Ghana's Ministry of Environment, Science, Technology and Innovation and the EPA, the promised facility and project combines health, environment, and economic goals. It aims to provide "in-depth retraining to enable proper recycling" and to "ensure the value-chain management of e-waste and electrical equipment in an environmentally sound manner thus turning the current challenges of E-waste management in the country into green business opportunities for the people of Ghana." The Ghanaian state has promised $30 million to the development of this new facility, a green e-waste management facility that extends the scope and variety of other forms of "techno-optimism" emerging in Ghana (Droney 2015). Agbogbloshie, as a new postcolonial frontier of neoliberalism (Chalfin 2010), endures as both a site of political-economic vitality and a platform for envisioning "green" alternatives.

To turn Agbogbloshie into a place of economic pride is for many in Agbogbloshie a necessary step for making e-waste recycling process and scrap metal recovery there less toxic and the labor less hazardous. This facility has not materialized, so the present and future of this facility are unknown. Proponents of this project might see this as a serious decontamination effort or option for avoiding future contamination from e-waste burning, but the bill was also passed just before the transfer of presidential power in 2016. What is also uncertain and on the minds of the copper burners I research is to what extent Dagomba migrants and laborers, those who make up the majority of Agbogbloshie's scrap metal workers, will be employed by future managers of the facility. The past and present promises and prospects of de-contamination in Agbogbloshie, I would suggest, have zero momentum unless these deeper labor and social marginalization politics are a front-and-center concern. Even amidst recent celebrations in Ghana pointing to its status as the first African state to "officially launch guidelines for envi-ronmentally sound e-waste management" (Sustainable Recycling Industries 2018), it is far too early to celebrate. For one thing, it is ultimately up to leading officers in government agencies like the Ministry of Environment, Science, and Technology and the EPA to turn these "guidelines" into action-able changes. Also, even if these agency leaders do enforce these guidelines and transform the country's e-waste management process, the "stakeholder" group, as defined by officers who created the guidelines, does not include e-waste burners themselves. How those stakeholders are folded into the management plan will determine the failure or success of the planned e-waste management overhaul.

Other examples of e-waste management informed by techno-optimism can be found in the web documentary project titled "E-waste Republic." Developed with the support of the Innovation in Development Reporting Grant program of the European Journalism Centre and additional funding from the Bill and Melinda Gates Foundation, this project seeks to stimulate interest in the new age of e-waste optimism in Ghana. In particular, this project is one of other emerging programs in Agbogbloshie that are seeking both solutions and social acceptance. One such project is the Agbogbloshie Makerspace Platform (AMP), which is a makerspace-oriented project that has valued positive engagement with workers from its inception. The AMP is a "transnational youth-driven project to promote maker ecosystems in Africa, starting at Agbogbloshie." According to DK Osseo-Asare, a co-founder of AMP,

> Our agenda is collective action: join hands to prototype tools and co-create a hybrid digital–physical platform for recycling, making, sharing and trading. We maintain that to realize an innovative future for Africa, we must start from the ground up: mining what already works for models and methods, and deploying co-design within existing communities of makers, across class, religious and tribal strata (Osseo-Asare, n.d.)

Based on several site visits to the AMP facility, which is located at the main entrance to the Agbogbloshie scrap metal market, I get the feeling that the facility mostly provides a location for workers to socialize and escape the sun or is a meeting spot for the occasional visit from international student groups touring Agbogbloshie to experience one of the world's most notorious e-waste dumpsites. But the vision of the AMP endures, and its co-founder is a makerspace enthusiast and determined techno-optimist driven to create alternative tech-based work opportunities for Agbogbloshie's e-waste laborers:

> A lot of times what people miss is that Agbogbloshie is a node in a network. It's not just the activities happening within the boundaries of Agbogbloshie but also its vast reach. Some of the people there are collecting scrap from as far away as Ivory Coast and Burkina Faso. They are actually bringing end-of-life equipment to this site, and there is a strategic collection of components, both in electronics and the automotive side. In that sense,

Agbogbloshie is a supplier. That's where we come in with the Agbogbloshie Makerspace Platform. We say, how can we "green" some of these very crude recycling practices? But equally, how can we leverage all of the materials that are here—and, in a sense, the expertise that's here—and transform it to a network for digital fabrication and distributed manufacturing. There's a lot of young people in Ghana and Africa. Where are their jobs going to come from in the future? We all know the future is going to be dominated by essentially robots and things that are electronic. We figure the best way to start educating a workforce is to learn from existing technology—learn how to fix things and innovate on top of that. (Osseo-Asare 2016)

Efforts to fix and innovate e-waste recycling in Agbogbloshie have been primarily anchored to economic development and environmental decontamination logics. While the AMP seems to be the most progressive of these interventions, or at least the one that seems to foreground the improvement of economic livelihoods, even the AMP approach overlooks how many scrap metal workers in Agbogbloshie continue to use "crude recycling practices," like wire burning. This is a matter of fact and concern that exposes site-specific pyropolitics of "green" techno-optimistic interventions that are largely informed by global environmental health science and reporting. According to a recent global health study, a global health intervention in Agbogbloshie has much to consider that goes well beyond better management or processing of e-waste. According to Yu et al. (2017, 96), "Generally, the workers did not perceive their work process and work environment in a positive light and might be amenable to informed decision-making on worker health with an increase in education and awareness creation about environmental pollution and the health risks associated with informal e-waste recycling." The report goes on to add that "agencies or organizations that are interested in implementing social intervention programs aimed at improving the livelihoods of workers should not only consider sustainable technologies to improve the safety and well-being of the workers, but should also provide opportunities for environmental and public health action, which engages and involves the e-waste workers directly in decision-making with support to learn other trades, training, and upgrade of skills appropriate for carrying out more environmentally sustainable recycling activities."

Between the work of the AMP, Pure Earth, the proposed SGS facility, and the various global health research teams working in Agbogbloshie, it is clear that Agbogbloshie has become a "node in a network" of various environmental health science studies and a node of e-waste decontamination and risk mitigation efforts. Moreover, it has become a node in a network of desperate techno-optimistic interventions that seem to regularly boom and bust and risk becoming as obsolete as the electronic discard showing up and accumulating in Agbogbloshie. But despite the ground-level failures, we can still learn from possibility and embrace the possible shifts to e-waste studies on the horizon, especially e-waste studies that engage with a subject matter that seems to, at times, defy representation.

Possibility of a Just E-Waste Transition in Ghana?

On November 17, 2017, an international group of e-waste researchers and policymakers joined together for the "E-Waste: Prevention Intervention Strategies Meeting" in New Delhi, India. The workshop was convened and financially supported by the US National Institute of Environmental Health Sciences (NIEHS), in collaboration with the Public Health Foundation of India and the Pacific Basin Consortium at the East–West Center. The meeting was convened largely in response to international environmental health studies focusing on health outcomes related to e-waste exposure, but what really kick-started the global conversation was a systematic review led by the World Health Organization (WHO) and its collaborating centers. This initial review showed that increases in spontaneous abortions, stillbirths, and premature births and reduced birth weights and birth lengths were as-sociated with exposures to e-waste. Additionally, it was found that direct and indirect exposures were a threat to human health, especially to vulnerable groups such as fetuses, children, pregnant women, the disabled, and workers in the informal sector. The WHO determined that these vulnerable groups needed specific protections, especially since the majority of global e-waste recycling is done informally, mostly by migrant workers using techniques such as burning with little or no safeguards in place for human and envi-ronmental health protection. The meeting held in 2017 stems from previous workshops convened in Depok, Indonesia, and Geneva, Switzerland, and a recent Global Environmental Health webinar sponsored by the NIEHS.[10]

---

[10] I participated in an NIEHS webinar on the relationship between e-waste and environmental health in 2018, which did include a research scientist from the University of Ghana. This expert shared little new information on the epidemiological research being conducted among a selection of Agbogbloshie workers.

The goal of the 2017 meeting was to work with international partners to develop prevention and intervention strategies to reduce exposures to e-waste to humans and the environment and to provide the community sector with tools that can allow the development of better surveillance, monitoring, and technologies with which to build capacity for better diagnosis and prevention and risk communication to workers and their families. In the summary report of the meeting it was noted that all e-waste interventions going forward should endorse "the following basic concepts": "No open, uncontrolled burning, such as wire burning to extract metals"; "No exposure of vulnerable persons to dismantling and recycling activities, especially pregnant women and young children in particular"; and "No unsafe and environmentally unsound dismantling." What these recommendations can often overlook is the tremendous power of informal economies in waste management and recycling practices across Africa, where less than half of the waste generated is actually "formally" collected. According to a recent World Bank report, "Because of moderate formal collection rates, open dumping and burning are commonly pursued to eliminate remaining household waste" (Kaza et al. 2018, 79). This is yet another reason why e-waste workers in Ghana engage in e-wastelanding. They are the ones who actually fill the waste management labor gap. Their "informal" labor is actually necessary to effectively manage Ghana's urban waste problem.

All of this is to say that Agbogbloshie, beyond being a site of vibrant e-waste trades and toxic exposures, is also an environment of uncertain e-waste intervention and contentious urban governance. While it is certainly the case that "ethnographic fieldwork makes one acutely aware of why the organizational and spatial dispersion of the state matters" (Gupta and Sharma 2006, 291), e-waste ethnography can also help us better understand the various social, environmental, and economic life politics informing the intervention environment that Agbogbloshie has become. By attending to experiences of socioeconomic precarity and environmental health suffering, we can more fully comprehend how and why these narratives diverge from the desires and logics of "green" techno-optimistic interventions inspired by what Howe and Boyer (2020, 2) might call the "verdant optimism" of e-waste management solutions, "a weird optimism [that] rings out, promising a sustainable way of doing more of the same . . . the same impulse to growth and ballooning wealth . . . as green only appears to capital and its agents in the monochrome of profit and accumulation."

Alongside these emerging solution-based and e-waste economization maneuvers, a turn to e-waste ethnography helps us translate and better grasp the place of cultural vitality in sites and situations of extreme e-waste pollution and residual neoliberalization. In Agbogbloshie we are confronted with a particular "sociotechnical imagination" whereby the dominant storyline of the e-waste problem there is trapped in a narrative purporting that e-waste is primarily a waste management and recycling problem. It is likely no surprise that this is precisely the circulating narrative that informs most e-waste studies. As Lepawsky recently put it, this has important consequences: "The character, distribution, and scale of harms and benefits arising from global flows of e-waste cannot be sufficiently understood if the dominant story line about waste dumping remains the unquestioned starting point for specifying discarded electronics as a waste management problem" (2018, 90). As I have attempted to highlight, what is also difficult to see and understand amidst this dominant story line are the underlying social politics of optimistic global environmental health interventions in zones of electronic discard. An ongoing challenge is to rethink the e-waste problem by adding nuance and other burning matters of attention to the storyline. This can and should involve attending to how understanding transboundary e-waste flows involves, on some level, the power and place of global Black ecologies (Hare 1970; Roane and Hosbey 2019). This latter issue is one I have spent little time exploring, but I think it matters to any critical discussion of future e-waste justice politics in Ghana. E-wastelanding is entangled in e-waste racialization, and e-waste politics in Ghana are fully entangled in global Black ecologies. These e-waste histories and social facts surely need to be told and acted on. As noted in the Introduction, "Just as race is a discursive technology with often deadly material effects, so too is wastelanding the process by which pollutability is materialized" (Voyles 2015, 15). Toxic e-wastelanding in Agbogbloshie, in this way, exposes and advances our understanding of the toxic materialization of racial politics of e-waste studies in Ghana and beyond.

The ways in which Agbogbloshie is rethought matter, and this rethinking effort calls for critically engaging the role of imagination in any e-waste intervention that seeks a just transition: "By turning to sociotechnical imaginaries, we can engage directly with the ways in which people's hopes and desires for the future—their sense of self and their passion for how things ought to be—get bound up with the hard stuff of past achievements" (Jasanoff 2015, 22). Agbogbloshie's e-waste interventions are made all the more complex and contentious because the narratives and imaginations of e-waste workers themselves are, if not fully excluded, dwarfed by celebratory anti-toxics interventions informed by neoliberal and techno-optimistic solutions to Ghana's e-waste problem. What matters is that e-waste interventions have been largely based on technology-based end-of-life management solutions, even though new approaches are on the rise. For example, it is exciting to learn that according to the International Labour Organization (2019) in Geneva, the range of e-waste interventions needs to be further developed. It is now recognized that the multiplicity of options for intervention derives from the multistakeholder landscape that makes up e-waste economics and management (e.g., industry reps, regulatory agencies, customs authorities, NGOs, etc.) (Lundgren 2012). How "sociotechnical imaginaries" inform even this wide-ranging intervention focus, I would argue, can provide a critical guide for emerging e-waste ethnography. Finally, all intervention in Ghana needs to be cognizant of not only decolonization itself but especially epistemic decolonization, as explored in the previous chapter. A more radical e-waste intervention in Agbogbloshie would be one that would avoid saturating intervention logics and projects with a techno-futures approach that only celebrates techno-scientific fixes to the e-waste problem. Such "progress" talk gets tricky because, as Droney (2015) reminds us, "science in Ghana is still infused with a sense of Black racial pride and a means of redressing racial disenfranchisement, and understanding this is key to understanding the global politics of science today" (p. 38). It is therefore critical to rethink and reimagine e-waste environmental health science and "expert" intervention along these lines as well. This seems especially critical if e-waste innovations and developments in Agbogbloshie are to be set in relation to discussions and critiques of science and technology for and in concert with "just transition" thinking (Newell and Mulvaney 2013; White 2020) and "cleantech" innovation and solution dreaming (Goldstein 2018) that is largely driven by an urbanized "tech-financial meritocracy" (Zukin 2020).

The Ghanaian state continues to contend that it is taking the e-waste management challenge and toxic disaster in Agbogbloshie seriously. This seriousness is being measured by occasional waves of "solutions-based" experimentation but "solutions" devoid of any actual or lasting infrastructure. Despite inflows of foreign direct investment and capital investments in public–private partnerships, the promised "state-of-the-art" e-waste recycling facility has still not come to fruition; but this has not stopped the momentum of political discourse. Future transition to a better and healthier environment in which to engage in e-waste recycling labor is the stated goal of current President Akufo-Addo, and it is important to note the attention to national security interests. In his words, "we are about to see an end to this global environmental challenge in Ghana, that is fast becoming a national security threat to most governments on the continent" (Akufo-Addo, 2018). The "end" of the problem is also framed within a discourse of jobs and job creation, with President Akufo-Addo claiming the new facility and network of proposed e-waste collection centers throughout the country will create over 22,000 "self-sustaining" jobs for Ghana's youth.[11] In addition to creating a nationwide and interagency e-waste fund to finance this transition to a healthier and environmentally friendly e-waste management program, Akufo-Addo encouraged those engaged in the e-waste trade sector to get involved:

> I encourage members of the various scrap dealers' associations, across the country, also to take full advantage of this programme. *I appeal to them to halt the burning of electrical and electronic wastes, as it only pollutes our environment and the quality of air we breathe.* They have an opportunity, through this programme, to expand their scrap collection ventures into big businesses. (Akufo-Addo, 2016, my emphasis)

Aside from these e-waste infrastructure and "green" development promises coming from the president's cabinet, what is perhaps most promising in recent e-waste news in Ghana is the emerging effort to build e-waste recycling industry capacity in the Northern Region. This is positive news, especially since most e-waste workers I have interviewed for this project have told me that they would like to live, work, and raise their family in the north, where they are from.

In February 2019, it was reported that Caritas Ghana, a development NGO of the Ghana Catholic Bishops Conference, had a 2-day workshop on e-waste management for the Tamale Ecclesiastical Province, a province that includes Tamale, Navrongo-Bolgatanga, Wa, Yendi, and Damango. The workshop was part of a larger e-waste project inspired by Pope Francis' "Care for Our Common Home" mission and seeks to address the e-waste problem in Ghana by addressing e-waste impacts on environment, human health and livelihoods. The project, which is financially supported by Deutsche

---

[11] In fact, in May 2021, a youth-driven movement emerged in Ghana called the #FixTheCountry protest, which directly contests this chronic experience of joblessness.

Gesellshaft fur Internationale Zusammenarbeit in partnership with City Waste Recycling Ltd., aims to both train youth on the collection and dismantling of e-waste and establish a dismantling factory in the north that can sustainably recycle e-waste. Sharing his optimism for the new project, Samuel Zan Akologo, the executive secretary of Caritas Ghana, noted, "We don't want the young people in the north traveling to Accra and facing all the difficulties that they are faced with in the Agbogbloshie scrap market." He highlighted a recent survey, which found that about 90% of the youth who make the journey to Agbogbloshie to make minimal wages were northerners in search of "greener pastures and improved livelihoods." He went on to add "It's very dangerous to burn electronic waste since it pollutes the air with dangerous chemicals resulting in diseases such as cancer and other things" (Caritas Ghana, 2018)

What these concerns and insights further illustrate is the growing need for grassroots interventions that look far beyond a vision of help that focuses squarely on improving e-waste recycling and making metal extraction and trade in and beyond Agbogbloshie more efficient. Surely that vision is shortsighted or, at the very least, contains elements of obsolescence that keep Agbogbloshie trapped within "easy" and reductive narratives (Lepawsky 2019). Ironically, the fact that e-waste burning continues makes even the best-intentioned interventions seem pointless and even obsolete. Agbogbloshie, in this sense, exists at the intersection of obsolete things and obsolete interventions. It exists as an ongoing place with ongoing toxic labor and ongoing toxic exposures. Therefore, any serious overhaul of e-waste intervention logics and practices in postcolonial Ghana, any effort to govern the ongoing toxic e-waste inferno must reckon with complex cultural, economic, environmental, and health uncertainties of the "e-waste commodity frontier" (Knapp 2016).

The story of e-waste in Agbogbloshie is now fully enwrapped in the pandemic uncertainties of COVID-19, a virus that has ignited global lockdowns and social distancing and mask-wearing mandates. At the time of this writing, Ghana had confirmed 95,236 cases and 794 deaths, based on limited testing (only 1,657 cases had been reported in the Northern Region compared to 52,419 cases in Accra). According to the Africa Centres for Disease Control, nearly 20 million Africans have received a vaccine through the COVAX program, and nearly 450,000 Ghanaians have already received their first dose of the AstraZeneca vaccine (Mwai 2021). Amidst these global vaccine rollouts, it remains unclear how Ghana's already stretched healthcare system will be able to respond to the threat of future coronavirus variants, especially amidst a mostly broken influenza-surveillance system across sub-Saharan Africa and because "serious flu cases are commonly conflated with malaria or just added to the 'acute respiratory infection' . . . grab bag" (Davis 2020, 56–57). As Bannister (2020) recently noted, while the pandemic has deepened the state of uncertainty and economic precarity, this is also a time of tough migration decisions and even postcolonial ridicule:

> Ghanaians joke about the "colonial virus" as a disease of the wealthy, and debate whether it is safer to remain in urban centres or return to family villages: the life of early lockdown unfolds with a comparable logic in many places. There is the sense of waiting for a wave to break. . . . Will Ghana's health system be overwhelmed? Will its accreted layers of old and new infrastructure, personnel, capacity and funding show resilience as case numbers begin to rise faster, if they do? Will Ghana's enduring situation at the periphery of the world economy contribute to worse outcomes and a lack of resources to fight coronavirus, as it has with some other diseases in the past? This marginality gave some early advantage, delaying the outbreak because of the country's relative isolation from international travel and trade, but this is changing with increased in-country transmission.

We will have to wait and see how the unraveling of "coronavirus capitalism" (Klein 2020b) will reshape the multibillion-dollar e-waste trade industry as a whole and its direct impacts on copper scrap trades and e-waste policies in Ghana in particular. For example, it appears that governments around the world are adopting economic "recovery" plans, with Big Tech and green technologies playing a central role.[12] The increasing role of tech-based foundations and private interests in pandemic "recovery" and global health research is also a matter of concern. "Fueled by more than $600 million in funding from the Bill & Melinda Gates Foundation—a virtually unheard-of sum for an academic research institute—the IHME [the University of Washington's Institute for Health Metrics and Evaluation] has outgrown and overwhelmed its peers, most notably the World Health Organization (WHO), which previously acted as the global authority for health estimates"

---

[12] As Pitt (2020) and Varoufakis (2021) have rightfully argued, contemporary capitalism has drifted toward "techno-fuedalism," and the COVID-19 pandemic has further fueled this political-economic transformation.

(Schwab 2020).[13] Alongside these coronavirus technocapital developments is the recent copper craze. Agencies like the Copper Development Association are keeping a close eye on "recovery" planning, development, and investment as they will drive global demand for copper (Copper Development Association 2020). Copper and coronavirus capitalism are merging, and this "new age of copper" (Taylor 2020) will likely influence markets in Ghana and across the Global South.

But the truth of the matter is that the coronavirus pandemic has "plunged the world into the worst economic crisis in the history of capital" (Harvey 2020). While COVID-19 has certainly upended healthcare systems and severely disrupted global supply chains,[14] how this will all reconfigure and reshape the lives of Ghana's e-waste laborers is unclear. How will the "colonial virus" intersect with the toxic colonialism of e-waste trades? What overlapping social and environmental health ruptures will emerge, and how will the global political ecology of e-waste generate new "uneven terrain[s] of development and power" (Moore 2011, 143)? How will "circular economy" enthusiasm, as an emerging solution to our planetary crisis of discard dystopia, actually better the lives of workers in Agbogbloshie? Addressing these question calls for first confronting the fact that "Any form of sustainable future (forget anything so grand as 'development') requires less for the prosperous among us, not more. Less stuff, less desire, less comfort, less convenience. Less of everything, really" (Hecht 2020). Less negativity, less toxic tropes of pollution and ruination. To really know e-waste life struggle, we will surely need to embrace the ongoing uncertainties and forms of plasticity that actually make e-waste ethnography a worthwhile method of education, intervention, and hope.

[13] It should also be noted that billionaire Bill Gates has taken on a personal interest in innovating human waste infrastructures to reduce diarrheal disease in Africa and the developing world. For more details on this fecal matter turn for this tech pioneer, see Davis Guggenheim's 2019 film *Inside Bill's Brain: Decoding Bill Gates.*

[14] Structural vulnerabilities in global supply chains are also made explicit with single events, such as the container ship stall-out that occurred in the Suez Canal in March 2021, which actually resulted in a semiconductor supply-chain crisis and concern for technocapitalists (Stahl 2021).

Conclusion

New Relations, Openings, and Burning Matters

With Agbogbloshie, it has been hard to move beyond the waste and pollution representation.

—Grace Abena Akese

Everywhere you look in Africa, people are constantly trying to put back together that which has been broken.

—Achille Mbembe (2020a)

This book has explored various burning truths and undercurrents of the e-waste pollution "crisis" in Ghana. Without real closure, I am stuck on the same questions that steered this project at its start: How can we know e-waste struggle and uncertainty in Ghana in a way that both focuses on and goes beyond toxic burning, global environmental health, and e-waste management and intervention logics? What kinds of e-waste truths emerge from e-waste ethnography? What might a decolonial e-waste intervention in Ghana look like? How can a pyropolitical ecology perspective complicate the "toxic" centering of e-waste labor? These burning questions linger. This book has struggled with these questions. In the shadows of these questions, I have tried to approach the subject of e-waste burning in Agbogbloshie by attending to the complex and precarious lives of workers moving in and out of an urban e-wasteland that is simultaneously a place of contamination, eviction, urban waste management, negative narration, e-waste mystification, e-waste epidemiology, and cultural livelihoods informed by kin and social networks, as well as chiefdom politics. My goal, after all, was never to treat e-waste burning as a narrow "problem" or an example of Ghana's postcolonial struggle to manage e-waste, nor was it my intention to treat Agbogbloshie like an airless container. Instead, Agbogbloshie is a fluid marketplace that is

Burning Matters. Peter C. Little, Oxford University Press. © Oxford University Press 2022.
DOI: 10.1093/oso/9780190934545.003.0008

under constant transformation, a situation that resists interpretational containment. Ethnography makes a certain situation out of a situation. It always makes things anew (Anand 2019). At the heart of the book, then, was an effort to attend to and *situate* e-waste burning in a way that showcased relations and matters of concern and struggle on the edges of e-waste pollution narratives. It turns out, e-waste burning, while certainly a practice that situates people within a global copper supply chain, is hardly all that defines the situation that these African e-waste workers find themselves in. Ghana is surely one of many "black holes of informational capitalism" (Castells 1997, 162), and Agbogbloshie is certainly more than a burning black hole of technocapital waste, a depleted landscape brutalized by the dual monstrosities of colonial and capitalist accumulation, mining, harvesting, and ruination.[1]

In the process of writing this book, it became clear that *Burning Matters* isn't really about "waste" after all. Instead, the focus became the actual vitality of copper wire burners in Ghana who experience a life between the hinterland and the urban margins and the ways in which these workers face countless forms of bodily struggle and socioeconomic uncertainty. While e-waste labor was a strong focus here, "waste" itself was not the front-and-center emphasis. There is a good reason for this. Detritus garbage, trash, recyclables, and rubbish are common foci of waster studies; but what discard studies try to do, what *Burning Matters* tries to do, is recenter and "include people, landscapes, futures, ways of life, and more," as Liboiron (2018a) puts it. What matters is how we make sure we leave "our objects of study open, because we are committed to a mode of inquiry (critical investigation of case studies) for a genre of things that are systematically devalued, cast out, erased, ignored, killed, removed, ruined, and otherwise cast in the negative" (Liboiron 2018a). My focus on the burning of electronic discard intended to open up the study to an analysis of relations set in motion by toxic fire, nongovernmental organization (NGO) intervention, urban governance, and environmental illness. Anchored to an ethnographic approach attuned to political ecology, environmental justice, and discard studies, I have intentionally left my pyropolitical ecology of e-waste in Ghana "open" to allow for theoretical vitality and ventilation. The term ought not be considered a suffocating

---

[1] As Povinelli (2019, 3) would have it, technocapital waste exposes deeper capital waste logics: "Capitalism depends on creating by destroying and then erasing the connections between the material wastes it leaves behind and the glimmering [and high-tech] oasis of privilege this waste affords."

trope but instead a terminological option or alternative "conceptual vector" (Millington and Lawhon 2019) for navigating emergent political ecologies of e-waste burning.

Doing fieldwork in Agbogbloshie is no doubt a toxic experience. I have experienced the headaches and coughing workers report, as well as the bodily exhaustion felt after only several hours of hanging out in this active scrap market. Just being in Agbogbloshie can be a hot and suffocating experience. It is a complicated cultural space, a challenging place to fully articulate and understand. In the face of this challenge, I have taken an open ethnographic and theoretical approach, a perspective aimed at finding fitting terms and concepts to describe this complex place of discard and cultural vitality. I experimented with developing a pyropolitical ecology of e-waste perspective to offer another terminological spark in a growing field of discard studies in general and critical e-waste studies in the Global South in particular. During fieldwork I was often consumed by the same smoke that became the target of "solutions-based" global environmental health interventions. In many ways, I came to know and understand Agbogbloshie not just as a place that one goes to and visits but as a place that gets on your clothes, hands, and feet and in your lungs. Everyday confrontation with toxic fire taught me that Agbogbloshie is a toxic atmosphere that gets embodied. Experiencing Agbogbloshie also complicates how I think about and story such a place. Ethnography taught me that the place and agency of e-waste in Agbogbloshie is also a burning matter, where e-waste is under fire, heat, and toxic incineration. These became more than metaphors for describing the pyropolitical ecology of e-waste recycling in postcolonial Ghana. In the process of paying attention to the e-pyropolitics unfolding in Agbogbloshie, I became curious about how e-waste burning is linked to broader, more global, waste incineration trends and politics. The burning of waste, including e-waste plastics, is a global problem that has become a primary focus of international organizations and think tanks, such as the United Nations University, the International Telecommunication Union, and the International Solid Waste Association, as well as global advocacy groups like the Global Alliance for Incinerator Alternatives. Burning waste, either for energy or to supply a global metal market, is a truly complex issue with a variety of political actors engaged in finding alternative and sustainable solutions to the multifarious problems associated with the incineration of modern things and discard. The "burning truth" (Minter 2016) of e-waste management problems and challenges is fast taking an African turn.

Africa has become a region of interest for many international agencies and organizations seeking to manage and control what seems, at times, an impossible toxic and global electronics production and e-waste recycling problem. According to one recent report, "Most African countries are now aware of and concerned with the dangers inherent to poor management of e-waste. However, the legal and infrastructural framework for achieving sound management still remains far from realised in the majority of countries" (Baldé et al. 2017, 60). The report also points out current "policy" challenges, explaining that most African states lack any formal or law-based e-waste management policies, with the exception of Uganda and Rwanda. Furthermore, while many African states have ratified the Basel Convention, only Madagascar, Kenya, and Ghana have drafted domestic laws specific to e-waste stream management, while South Africa, Zambia, Cameroon, and Nigeria are still working to get e-waste bills passed. Even in places like Nigeria and Ghana, where e-waste importation laws are apparently being enforced, those engaged in the informal e-waste sector have found clever ways to both import and receive "illegal" e-waste shipments through porous state boundaries. In Kenya, an African state that generates nearly 45,000 tons of e-waste annually, e-waste legislation is in draft form that will make it a requirement for electronics companies to have clear waste management plans and will prohibit the importation of e-wastes that don't have a clear end-of-life management plan.

Since 2016, Ghana has had in place an e-waste bill that prohibits imports and exports of e-waste. While enforcement is largely lacking, this bill does take a step toward creating a registration system for manufacturers, importers, and distributors of e-wastes or other discarded electrical equipment. The bill also focuses on "phasing out the inclusion of printed circuit boards in electronic equipment . . . as well as the establishment of an e-waste management fund to be achieved through payment of an advance eco-fund by manufacturers, importers, and distributors" (Baldé et al. 2017, 60). Following these and other government efforts to manage e-waste, many African states have begun to attempt to find ways to integrate the informal sector, a sector which dominates many e-waste trades and many recycling/circular economies. But these "integration" efforts face many challenges, namely that most of these countries have limited access to electronics "take-back" programs or lack appropriate infrastructure to effectively recycle electronic discard. Based on my experience in Ghana, even when legislative bills to govern e-waste appear, it is inefficient largely because the very process of "integrating" so-called formal and informal sectors involves integrating Ghanaians with different life experiences based on different cultural, economic, and educational privileges. Poor Dagomba working in Agbogbloshie don't share the same life experience of government officials and experts who develop e-waste management and integration plans and who have the social, political, and economic power to build e-waste and health intervention "capacity" more broadly (Tousignant 2018). These differences matter.

Many African states are currently receiving "advisory, technical, and financial support from several [United Nations] agencies, other development agencies, the private sector, and especially from the alliance of Original Equipment Manufacturers . . . in Africa" (Baldé et al. 2017, 60). For example, recently a consortium of international experts assisted Ghana in the development of a training manual for e-waste recyclers (Deutsche Gesellshaft fur Internationale Zusammenarbeit 2019).[2] Similar forms of international expertise and techno-optimistic solutions to e-waste management in Africa are emerging. This is generally good news. But it is always important to learn who actually benefits and who doesn't. Plans to "sustainably" manage e-waste in Ghana are generally unclear about how they intend to include poor Dagomba workers in Agbogbloshie and Old Fadama. Moreover, a plan for their *sustained* inclusion remains ambiguous. That uncertainty matters and calls for continued e-waste ethnographic attention, inflection and critique.

Agbogbloshie is a recognized site of challenge among the growing international community of e-waste management actors working on solutions and "best practices." But it is also a site of recognized intervention failure. The efforts of Pure Earth were given early praise, and community support and optimism seemed to have some momentum; but to date, the intervention has basically dissolved, and the infrastructure itself is in slow decay. The intervention has basically bottomed out. The solution to integrate and standardize e-waste recycling in Agbogbloshie has gone stagnant, despite ongoing efforts by organizations like Green Advocacy Ghana. This narrative of e-waste response failure is not unique to places like Agbogbloshie. Similar intervention efforts in Kenya, Uganda, and Tanzania have struggled to sustain projects and facilities due to "poor business management decisions" (Baldé et al. 2017). This has led to a more recent interest in establishing private recycling

---

[2] This manual utilized a framework developed by the German corporation and was meant to support the Ghanaian Ministry of Environment, Science, Technology and Innovation to "improve the conditions for sustainable management of e-waste in Ghana" (Deutsche Gesellshaft fur Internationale Zusammenarbeit 2019, 3).

plants and issuing public–private partnerships across the continent to effectively manage Africa's real, yet precarious, e-waste economy and pollution
crisis. This neoliberalization of e-waste management and circular economy
solutions will surely lead to more, not less, uncertainties for Ghana's e-waste
laborers. Furthermore, the risks of "technological disemployment" are now
real concerns for waste workers navigating an already precarious "new
economy" landscape (Pierce, Lawhon, and McReary 2020; Smith 2000).

In Agbogbloshie, I confronted more than weak environmental NGO
intervention and a lack of state support for Ghana's hard-working poor.
I learned about Dagomba living and "surviving eviction" (Menon-Sen and
Bhan 2008) and toxic labor in Ghana's urban margins yet also witnessed
"solutions" with no direct engagement with environmental justice politics,
despite the fact that emerging research and activism in Africa are grappling
with "the dispossession and toxic pollution involved in extractivism" (Cock
2015). I learned that toxic burning and copper extractivism in Agbogbloshie
are symbolically and politically linked to broader global waste and environmental justice struggles. Global supply-chain capitalism involves
sites or nodes of toxic technocapital that connect lives across Africa, from
Congolese miners extracting cobalt (Kara 2018) to tungsten miners in
Uganda (Lewis 2019) to those burning e-waste to extract copper in Ghana.
No matter their place and position on the African continent, these workers
face similar experiences of contamination, dispossession, and struggle. In
most cases, these urban e-waste laborers are similarly "dispossessed people
who . . . for the most part no longer practice agriculture" (Smith 2017, 296).
Moreover, global consumption of electronics and high-tech capitalism itself involve unjust injury, "even our collective planetary injury" (Murphy
2017, 140). Burning e-waste brings this injury to the surface and exposes
the bodily–environmental injustices of this toxic, yet life-supporting, labor.
Agbogbloshie exposes, flips, and distorts global narratives of injustice in a
way that further illustrates our "techno-precarious" times:

> It is tempting to think of all this in terms of an unequal relationship between
> the global north, as exporter of toxicity, and the global south, to which it
> is exported. This would be an over-simplification, but it cannot be denied
> that surplused populations across the globe are often the most precarious
> test subjects that serve the depletion economy. And the cardinal split of the
> globe doesn't quite cut it, as the proliferation of toxicity in depletion zones
> attests. (Precarity Lab 2020, 28)

As Grace Abena Akese and I have argued elsewhere (Akese and Little 2018), emerging politics from the Agbogbloshie case highlight important dimensions of Global South environmental justice that can also play a rejuvenating role in discard studies. We suggest that a more pluralist approach to (in)justice in Agbogbloshie is needed, an approach that opens e-waste justice politics up to terms and explications of injustice that go beyond the usual focus. So, while we note that "Like other e-waste hotspots in the global south, the claim of environmental injustice in Agbogbloshie finds its central referent in assertions that poor and marginalized populations in the global south are poisoned with toxic substances as they process the discards of rich countries in the global north" (Akese and Little 2018, 4), my research also highlights the need for an e-waste justice politics that also involve deeper engagement with the fact that e-waste is just as poisonous as the "discursive containment and pathology" (Simpson 2014) that make Agbogbloshie a certain stand-out place in our "toxic world" (Nading 2020). To merge e-waste justice and ethnography and to make the anthropology of e-waste in Ghana a more just endeavor requires admitting to the fact that White ethnographic representations of Agbogbloshie always run the risk of repeating and reproducing these same negative and toxic discursive falsehoods (Benjamin 2018).[3] What this e-waste ethnography can offer up is an angle on e-waste politics attuned to "the conditions of possibility" (Appel 2019, 5) that made e-waste economies and ecologies in Ghana possible at all.

There are certainly many reasons why an ethnographic synthesis of environmental justice, e-waste risk, and metal supply-chain studies is needed. Doing e-waste ethnography to help make sense of this synthesis in a postcolonial West African context is a matter of folding lots of concerns together and engaging in complex engagement with the dual commitments of ethnographic refusal and ethnographic insistence (Ortner 1995; Simpson 2007, 2014; McGranahan 2016; Appel 2019). Doing e-waste ethnography in Ghana has taught me that if indeed e-waste justice is to be realized at all, a recentering of the lives, role, and capacity of e-waste workers themselves is a necessary step. This redirection calls for a refusal of the usual negative Agbogbloshie storyline (Akese 2020). If nothing else, ethnographies of

---

[3] As Benjamin (2018) rightfully contends, "Vampirically, white vitality feeds on black demise— from the extraction of (re)productive slave labor to build the nation's wealth to the ongoing erection of prison complexes to resuscitate rural economies—in these ways and many more, white life and black death are inextricable. Racist structures not only produce, but *reproduce* whiteness, by resuscitating the myth of white innocence that inheres in the racial status quo. Racist systems are thereby reproductive systems."

e-waste in Africa and the Global South more generally can help us refuse falsehoods and avoid circulating mystification and the tendency to anchor e-waste politics across the continent to a desperate and negative "urban" slum crisis narrative.[4] This narrative might help capture the actual toxic environmental realities of global e-waste trades and help steer solution-based environmental NGOs, but it most certainly produces monochromatic framings of e-waste struggle and lived experience. The fact is, other burning truths and global uncertainties matter.

For sure, the global COVID-19 pandemic calls for a renewed e-waste ethnography attuned to global economic rifts and drifts. As anthropologist Arjun Appadurai recently put it,

> Just as COVID-19 is a virus with global qualities, globalization is itself viral. But all viruses evolve, and globalization is no different. The nation-states of the world have had highly uneven success in their response to COVID-19, and that is primarily because the architecture of the system of nation-states is not well suited to an age of problems without national boundaries. Since globalization is here to stay, it is the system of nation-states that might well be forced to change, in ways that some might welcome, and others will resist. That battle will outlast the story of COVID-19. (Appadurai 2020)

For example, it is currently unclear how Ghana and its scrap metal economy and e-waste recycling sector will change as a result of the ratification of the African Continental Free Trade Area agreement (AfCFTA), introduced in January 2012 at the 18th African Union Assembly to provide a comprehensive framework for attaining regional development across Africa by creating a single continental market for goods, services, and investments projected to involve more than 1.2 billion people with a combined gross domestic product (GDP) of more than US$3.4 trillion.

In May 2019, AfCFTA finally went into force, and Ghana projects an increase in the importation of raw materials from the region, a scaling up of manufacturing sectors, and increased regional trade. How this will directly or indirectly impact scrap metal trades is unclear. But with 2020 being an election year, it is expected that the government will overspend and undertax, a fiscal situation that will only increase Ghana's crippling national

---

[4] For an overview of how the concept and practice of ethnographic refusal figure in recent discard studies, see Zahara (2016).

debt situation. Currently, over 60% of Ghana's total wealth (a GDP of roughly $65 billion) is in the form of debt. In the language of the World Bank, Ghana is now a "highly indebted poor country"; and increased foreign participation in Ghana's debt exposes the country to high-risk global market swings and fluctuating foreign exchange rates. Under these conditions, it is very likely that no matter how much capital investment occurs in Ghana, no matter the intentions of foreign direct investments and public–private partnerships, most Ghanaians, especially marginalized Dagomba farmers and e-waste laborers, will only experience further uncertainties in response to global neoliberal austerity measures in the near and uncertain future (Blyth 2013).

Today, and amidst a hyper–debt crisis that won't likely be reversed with "development" and infrastructural promises coming from the re-election of Nana Akufo-Addo (New Patriotic Party) in 2020 presidential elections,[5] the Ghanaian state is politically engaged in active petrocapitalism (Mitchell 2011; Chalfin 2019). Offshore oil extraction along the Ghana–Côte d'Ivoire border continues to dominate much of the country's capital investments, but this "development" sector in Ghana is heavily dominated by a powerful British petro corporation (Tullow Oil), making mega-profits that Ghanaians don't see. During fieldwork in July 2017, one engineer working in Ghana's largest offshore oil project, called the "Jubilee" oil field project, explained that profits are being celebrated but that infrastructural risks are a concern:

> The Jubilee oil field just celebrated its 200 million barrel mark. That is around $20 billion. It achieved that mark in about seven years, when the oil field was first tapped. On average it produces about 150,000 barrels a day. They expect the field to produce a total of 500 million barrels. We are having drilling problems that have raised concerns and target production and revenues are not being reached. This has been a problem. (Interview, July 22, 2016)

The underlying story of "development" in Ghana is that private investment interests will surely outweigh social/community interests, which was exactly what happened after the privatization of the gold mining sector in Ghana in the mid-1980s.[6] Taking that same path in the e-waste recycling sector will surely mimic and sustain these forms of social and economic

---

[5] Members of the opposition party, the National Democratic Congress, have opposed the election results (Kokutse 2020), with some even burning tires as a symbolic gesture of dissatisfaction.

[6] As Ferguson (2006, 36) notes,

marginalization. As one Ghanaian friend told me, "You don't create a monster to fix a problem that can turn around and become a problem." Whether burning fossil fuels or burning e-waste, too many across Africa are themselves left to burn in the ashes of this monstrous debt-economy "development" machine that is repeatedly touted as the ultimate problem-solver. But the fact is, Ghana's e-waste "problem" involves a much deeper political economy and ecology of struggle. It is indeed a problem and struggle historically fastened to global Black ecologies and geographies of commodity extractivism, exploitation, racism, and dispossession. *Burning Matters* only scratches the surface of these deeply layered ecologies. It is simply a start to the longer project of synthesizing emerging scholarship on "Black ecologies" (Hare 1970; Roane and Hosbey 2019) with critical studies of global e-waste and e-wastelanding. This merger can guide emerging anthropologies of "ruination science" (Ureta 2021) more generally and help build e-waste anthropology futures rooted in more radical and decolonial political ecologies in particular (Jobson 2019).[7] While this book certainly falls short of recent methodological decolonization callings and forms of decolonial praxis (Murphy 2020; Mignolo and Walsh 2018; Walsh 2020)—including those aiming to decolonize White Anthropocene narratives and mantras of earthly destruction (Yusoff 2018)—it has tried to keep alive the sustained calling of decoloniality so clearly captured by Fanon (1952): "O my body, makes me always a man who questions!"

At its best, e-waste ethnography in Agbogbloshie can help us ask better questions that may even help us make sense of the multifarious ways in which e-waste is more than toxic e-waste ruination. Agbogbloshie is not just a toxic pyrocumulous cloud or technocapital deathscape. Instead, this e-waste ethnography exposed ongoing relations and matters of concern, uncertainty, and socio-environmental life happening on this "blasted" landscape (Tsing 2014). We need this perspective, these "situated knowledges"

The privatization of the gold mines . . . combined with generous tax incentives, has done just what it is supposed to do: bring in large amounts of private investment. Thanks to such investment, Ghana's gold industry has undergone a massive transformation since the mid-1980s. . . . Most important (and in contrast to earlier, more-labor intensive mining ventures), there has been little creation of employment for Ghanaians because of the "highly capital intensive nature of modern surface mining techniques." (World Bank 2003, 23)

[7] This move might involve seriously considering the pyropolitical in anthropology itself: "anthropology cannot presume a coherent human subject as a point of departure but nonetheless must adopt a new humanism as its political horizon. We may begin by letting anthropology burn" (Jobson 2019, 267).

(Haraway 1988), not because they necessarily correct misunderstandings of our global e-waste predicament but because ethnography helps open up e-waste to wider experiences and relations of struggle and precarity. My *pyropolitical ecology of e-wastelanding* approach was never bent on providing a more exact science of e-waste conflict but instead aimed to expose persistent *relations* (Strathern 2020) of concern that matter to e-waste studies anywhere. These relations spread and drift like a pyrocumulous cloud. Engaging a situated anthropology and political ecology of e-waste taught me that certain toxic truths and struggles drift and link to truths and justice struggle beyond Agbogbloshie and Ghana. If e-waste matters at all, we must reckon with the fact that, alongside the ongoing chronicling of toxic suffering, Black ecologies and Black Lives Matter everywhere.[8] From anti-racism movements, anti-pipeline movements, and environmental justice movements to reparation and abolition actions, all somehow "reject the slow deaths and fast violence everywhere" (Sze 2020, 102). Neoliberal power and state violence choke and dispossess people (especially people in poverty and people of color) everywhere. Zombie fossil fuel extraction, climate disruptions, and planetary warming stress out life everywhere. Now humans everywhere are witnessing the hyper-mess of technocapital depletion and the uneven distribution of toxic plunder. At its core, this is the raw power of technocapitalism, the dirty supply chains of what Povinelli (2018) terms "geontopower," where the ravagement of lands and bodies is alive and sustained by enduring forms of colonialism and capitalism in both the Global North and South. Furthermore, as Hecht (2020, 1) puts it, "We are turning the world inside-out. Massive mining operations rip into rock, unearthing lithium, coltan and hundreds of other minerals to feed our gargantuan appetite for electronic stuff."

In these messy, uneasy, and uncertain times, neither melting ice sheets, electronic discard, nor toxic smoke and flesh ought to be treated as inanimate, depolitical, and lifeless objects. These unsettling and fraught times call on us to avoid the lure of settling for narratives firmly fixed to the ruinous and eco-nihilistic dimensions of e-waste. E-waste is far from dead matter or material devoid of life. Surely e-waste toxicity matters, but we must continue to find ways to dodge the trappings of universal negatives that steer our attention away from understandings of e-waste that "actually contribute to livelihood and wellbeing" (Amankwaa 2014, 13). Toxic burning and

---

[8] See David Pellow (2016) for a discussion of the ways in which the Black Lives Matter movements overlap with critical environmental justice.

urban pollution may have put Agbogbloshie on the global e-waste dumping map, but if we truly wish to know and learn from the social, environmental, and political complexities and uncertainties of e-waste lived experience in Ghana, or anywhere in the Global South where e-waste fires blaze, then attending to other burning matters of life and labor is a must. What matters is what one does with what is shown and what is seen, heard, and witnessed. It is about confronting e-waste vitalism and precarity in the "burning of the world" (Mbembe 2020a, 2020b) and keeping a critical eye on the life and death, the late industrial madness and toxic planetarity of a technocapitalism that endlessly extracts wealth while simultaneously plundering and harming living bodies and landscapes (Precarity Lab 2020). Ethnography is one way to engage, rekindle, and repair life on the edges of e-wastelanding; but we mustn't do this to just challenge the pyrocumulous traps and politics of dominant e-waste storylines. E-waste ethnography must embrace the challenge of generating new and more righteous clouds of e-waste critique and theory to resensitize how we know e-waste and to build critical solidarity, hope, and even a common optimism in these burning and uneasy times.

References

Abrokwa-Ampadu, R. 1984. *The Volta River Hydro-Electric Project in Ghana; Hydro-Environmental Indices: A Review and Evaluation of Their Use in the Assessment of the Environmental Impacts of Water Projects*. Paris: UNESCO.

Adams, Vincanne, ed. 2016. *Metrics: What Counts in Global Health*. Durham, NC, and London: Duke University Press.

Accra Metropolitan Assembly. Accessed October 2018. ama.gov.gh.

Adelman, Jeremy. 2020. "The End of Landscape? Photography, Globalization and Climate Change." *New Left Review* 126. https://newleftreview.org/issues/ii126/articles/jeremy-adelman-the-end-of-landscape.

Adogla-Bessa, Delali. 2018. "Stop Playing Politics with Agbogbloshie Demolition—AMA Boss to Nii Lante." August 29. https://citinewsroom.com/2018/08/stop-playing-politics-with-agbogbloshie-demolition-ama-boss-to-nii-lante/.

Afenah, Afia. 2012. "Engineering a Millennium City in Accra, Ghana: The Old Fadama Intractable Issue." *Urban Forum* 23: 527–540.

Agamben, Giorgio. 1998. *Homo Sacer: Sovereign Power and Bare Life*. Translated by Daniel Heller-Roazen. Stanford, CA: Stanford University Press.

Agyepong, Heather. 2014. "'The Gaze on Agbogbloshie': Misrepresentation at Ghana's E-Waste Dumpsite." *OkayAfrica*, October 2.

Akese, Grace Abena. 2014. "Price Realization for Electronic Waste (E-Waste) in Accra, Ghana." Master's thesis, Memorial University of Newfoundland.

Akese, Grace Abena. 2020. "Researching Agbogbloshie: A Reflection on Refusals in Fieldwork Encounters." *Feministisches Geo-RundMail* 83: 52–55.

Akese, Grace Abena, and Peter C. Little. 2018. "Electronic Waste and the Environmental Justice Challenge in Agbogbloshie." *Environmental Justice* 11 (2): 77–83.

Akufo-Addo, Nana. 2016. Quoted in "E-Waste Recycling Facility to Be Constructed in Agbogbloshie." August 29. Communications Bureau, Office of the Presidency, Republic of Ghana. https://presidency.gov.gh/index.php/briefing-room/news-style-2/800-e-waste-recycling-facility-to-be-constructed-in-agbogbloshie-president-akufo-addo.

Alexander, Catherine, and Joshua Reno, eds. 2012. *Economies of Recycling: The Global Transformation of Materials, Values and Social Relations*. London: Zed Books.

Amankwaa, E. F. 2014. "E-Waste Livelihoods, Environment and Health Risks: Unpacking the Connections in Ghana." *West African Journal of Applied Ecology* 22 (2): 1–15.

Amankwaa, Ebenezer Forkuo, Kwame A. Adovor Tsikudo, and Jay A. Bowman. 2017. "'Away' Is a Place: The Impact of Electronic Waste Recycling on Blood Lead Levels in Ghana." *Science of the Total Environment* 601–602: 1566–1574.

Amoako, Clifford. 2016. "Brutal Presence or Convenient Absence: The Role of the State in the Politics of Flooding in Informal Accra, Ghana." *Geoforum* 77: 5–16.

Amoako, Clifford, and Daniel Kweku Baah Inkoom. 2018. "The Production of Flood Vulnerability in Accra, Ghana: Re-thinking Flooding and Informal Urbanization." *Urban Studies* 55 (13): 2903–2922.

Amoako, Clifford, and E. Frimpong Boamah. 2015. "The Three-Dimensional Causes of Flooding in Accra, Ghana." *International Journal of Urban Sustainable Development* 7 (1): 109–129.

Anand, Pandian. 2019. *A Possible Anthropology: Methods for Uneasy Times.* Durham, NC: Duke University Press.

Anarfi, J. K., Kofi Awusabo-Asare, and Nicholas N. N. Nuamah. 2000. *Push and Pull Factors of International Migration. Country Report, Ghana.* Luxembourg: Statistical Office of the European Communities.

Andrews, Kehinde. 2016. "Black Is a Country: Building Solidarity Across Borders." *World Policy Journal* 33 (1): 15–19.

Antoine, Adrien. 1985. "The Politics of Rice Farming in Dagbon, 1972–1979." PhD diss., School of Oriental and African Studies, University of London.

Appadurai, Arjun. 2020. "Coronavirus Won't Kill Globalization. But It Will Look Different After the Pandemic." *Time*, May 19. https://time.com/5838751/globalization-coronavirus/.

Appel, Hanna, Nikhil Anand, and Akhil Gupta. 2018. "Introduction: Temporality, Politics, and the Promise of Infrastructure." In *The Promise of Infrastructure*, edited by Nikhil Anand, Akhil Gupta, and Hannah Appel, 1–40. Durham: Duke University Press.

Appel, Hannah. 2019. *The Licit Life of Capitalism: US Oil in Equatorial Guinea.* Durham, NC, and London: Duke University Press.

Arboleda, Martín. 2018. "Planetary Mine as an Archaeology of Labor Futures." *Harvard Design Magazine* 46: 86–93.

Arboleda, Martín. 2020. *Planetary Mine: Territories of Extraction in the Fourth Machine Age.* London: Verso.

Archibald, Sally, A. Carla Staver, and Simon A. Levin. 2012. "Evolution of Human-Driven Fire Regimes in Africa." *Proceedings of the National Academy of Sciences of the United States of America* 109 (3): 847–852.

Armah, Frederick Ato, Reginald Quansah, David Oscar Yawson, and Luqman Abdul Kadir. 2019. "Assessment of Self-Reported Adverse Health Outcomes of Electronic Waste Workers Exposed to Xenobiotics in Ghana." *Environmental Justice* 12 (2): 69–84.

Armiero, Marco. 2021. *Wasteocene: Stories from the Global Dump.* Cambridge: Cambridge University Press.

Aryeetey, Ernest, and Finn Tarp. 2000. "Structural Adjustment & After: Which Way Forward?" In *Economic Reforms in Ghana: The Miracle and the Mirage*, edited by Ernest Aryeetey, Jane Harrigan, and Machiko Nissanke, 344–365. Oxford: James Currey.

Asante, K. A., S. Adu-Kumi, and K. Nakahiro. 2011. "Human Exposures to PCBs, PBDFs and HBCDs in Ghana: Temporal Variation, Sources of Exposure and Estimation of Daily Intakes by Infants." *Environment International* 37: 921–928.

Asante, K. A., T. Agusa, and C. A. Biney. 2012. "Multi-Trace Element Levels and Arsenic Speciation in Urine of E-Waste Workers from Agbogbloshie, Accra in Ghana." *Science of the Total Environment* 424: 63–73.

Asante, Kwadwo Ansong, John A. Pwamang, Yaw Amoyaw-Osei, and Joseph Addo Ampofo. 2016. "E-Waste Interventions in Ghana." *Reviews on Environmental Health* 31 (1): 145–148.

Assefa, Hizkias. 2000. "Coexistence and Reconciliation in the Northern Region of Ghana." In *Reconciliation, Justice, and Coexistence: Theory & Practice*, edited by Mohammed Abu-Nimer, 165–186. Lanham, MD: Lexington Books.

Austin, G. 1987. "The Emergence of Capitalsit Relations in South Asante Cocoa-Farming." *Journal of African History* 28: 259–279.

Auyero, Javier, and Déborah Alejandra Swistun. 2009. *Flammable: Environmental Suffering in an Argentine Shantytown*. New York: Oxford University Press.

Awumbila, M., J. K. Teye, and J. A. Taro. 2017. "Social Networks, Migration Trajectories and Livelihood Strategies of Migrant Domestic and Construction Workers in Accra, Ghana." *Journal of Asian and African Studies* 52 (7): 982–996.

Azmanova, Albena. 2020. *Capitalism on Edge: How Fighting Precarity Can Achieve Radical Change Without Crisis or Utopia*. New York: Columbia University Press.

Baabereyir, Anthony, Sarah Jewitt, and Sarah O'Hara. 2012. "Dumping on the Poor: The Ecological Distribution of Accra's Solid Waste Burden." *Environment and Planning A* 44 (2): 297–314.

Baldé, C. P., V. Forti, V. Gray, R. Kuehr, and P. Stegmann. 2017. *The Global E-Waste Monitor 2017*. Bonn, Germany: United Nations University; Geneva: International Telecommunication Union; Vienna: International Solid Waste Association.

Baldwin, James. (1962) 1993. *The Fire Next Time*. New York: Vintage International.

Bannister, David. 2020. "Futures and the Past in Ghana's Universal Health Coverage Infrastructure." *Somatosphere*, April 27. http://somatosphere.net/2020/universal-health-coverage-ghana.html/.

Bardsley, Phoebe. 2019. "Why Ghana's Biggest Dump Site Should Be Seen as an Innovation Hub." *African Center for Economic Transformation* (blog), July 31. https://acetforafrica.org/media/blogs/why-ghanas-biggest-dump-site-should-be-seen-as-an-innovation-hub/.

Basel Action Network. 2002. *Exporting Harm: The High-Tech Trashing of Asia*. Seattle, WA: Basel Action Network.

Basel Action Network. 2004. Comment on the Analysis of Issues Related to Annex VII. http://archive.ban.org.

Basel Action Network. 2005. *The Digital Dump: Exporting Re-use and Abuse to Africa*. Seattle, WA: Basel Action Network.

Basel Action Network. 2018. *Holes in the Circular Economy: WEEE Leakage from Europe*. Seattle, WA: Basel Action Network.

Basel Action Network. n.d. Mission. https://www.ban.org/mission.

Basel Convention. n.d. Accessed February 28. http://www.basel.int/Implementation/Ewaste/EwasteinNigerandSwaziland/EwasteinAfrica/Overview/tabid/2546/Default.aspx.

Beck, Ulrich. 2006. "Living in the World Risk Society." *Economy and Society* 35 (4): 329–345.

Behar, Ruth. 1997. *The Vulnerable Observer: Anthropology That Breaks Your Heart*. Boston: Beacon Press.

Benanav, Aaron. 2020. *Automation and the Future of Work*. London: Verso.

Bening, R. B. 1990. *A History of Education in Northern Ghana, 1907–1976*. Accra: Ghana Universities Press.

Benjamin, Ruha. 2018. "Black AfterLives Matter: Cultivating Kinfulness as Reproductive Justice." In *Making Kin not Population*, edited by Adele Clark and Donna Haraway, 41–66. Chicago: Prickly Paradigm Press.

Benjamin, Ruha, ed. 2019a. *Captivating Technology: Race, Carceral Technoscience, and Liberatory Imagination in Everyday Life*. Durham, NC: Duke University Press.

Benjamin, Ruha. 2019b. *Race After Technology*. Cambridge: Polity Press.

Bennet, Jane. 2010. *Vibrant Matter: A Political Ecology of Things*. Durham, NC: Duke University Press.

Bennett, Tony. 1993. "The Exhibitionary Complex." In *Culture/Power/History*, edited by N. Dirks, G. Eley, and S. Ortner, 123–154. Princeton, NJ: Princeton University Press.

Berry, Sara. 2009. "Property, Authority and Citizenship: Land Claims, Politics and the Dynamics of Social Division in West Africa." *Development and Change* 40 (1): 23–45.

Bhattacharyya, Gargi. 2018. *Rethinking Racial Capitalism: Questions of Reproduction and Survival*. London: Rowman and Littlefield.

Biehl, João, Byron Good, and Arthur Kleinman, eds. 2007. *Subjectivity: Ethnographic Investigations*. Berkeley: University of California Press.

Biehl, João, and Adriana Petryna. 2013. "Critical Global Health." In *When People Come First: Critical Studies in Global Health*, edited by J. Biehl and A. Petryna, 1–22. Princeton, NJ, and Oxford: Princeton University Press.

Black, Jan Knippers. 1999. *Inequity in the Global Village: Recycled Rhetoric and Disposable People*. Bloomfield, CT: Kumarian Press.

Black, Richard. 2004. "E-Waste Rules Still Being Flauted." *BBC News*. March 19, http://news.bbc.co.uk/2/hi/science/nature/3549763.stm.

Blakeley, Grace. 2020. *The Corona Crash: How the Pandemic Will Change Capitalism*. London: Verso.

Bleiker, Roland, and Amy Kay. 2007. "Representing HIV/AIDS in Africa: Pluralist Photography and Local Empowerment." *International Studies Quarterly* 51: 139–163.

Blyth, Mark. 2013. *Austerity: The History of a Dangerous Idea*. London: Oxford University Press.

Boadi, Kwasi, and Markku Kuitunen. 2002. "Urban Waste Pollution in the Korle Lagoon, Accra, Ghana." *The Environmentalist* 22: 301–309.

Boafo-Arthur, Kwame. 2007. "A Decade of Liberalism in Perspective." In *Ghana: One Decade of the Liberal State*, edited by Kwame Boafo-Arthur, 1–20. Dakar: CODESRIA Books.

Bob-Milliar, George M. 2016. "Chieftancy, Diaspora, and Development." In *The Ghana Reader: History, Culture, Politics*, edited by Kwasi Konadu and Clifford C. Campbell, 390–395. Durham, NC: Duke University Press.

Bogner, Artur. 2016. "The 1994 Civil War in Northern Ghana." In *The Ghana Reader: History, Culture, Politics*, edited by Kwasi Konadu and Clifford C. Campbell, 341–343. Durham, NC: Duke University Press.

Bond, David. 2017. "Oil in the Caribbean: Refineries, Mangroves, and the Negative Ecologies of Crude Oil." *Comparative Studies in Society and History* 59 (3): 600–628.

Bond, Patrick. 2006. *Looting Africa: The Economics of Exploitation*. London and New York: Zed Books.

Borowy, Iris. 2019. "Editorial Introduction to the Special Collection 'Development of Waste-Development as Waste.'" *Worldwide Waste: Journal of Interdiscplinary Studies* 2 (1): 12.

Boudia, Soraya, Angela N. H. Creager, Scott Frickel, Emmanuel Henry, Nathalie Jas, Carsten Reinhardt, and Jody A. Roberts. 2018. "Residues: Rethinking Chemical Environments." *Engaging Science, Technology, and Society* 4: 165–178.

Bourgoing, Robert. 2002. "Ghana: The Nightmare Lagoons." Accessed April 2, 2018, http://www.bourgoing.com/ghana.htm.

Boyer, Dominic. 2015. "Anthropology Electric." *Cultural Anthropology* 30 (4): 531–539.

Brempong, Nana Arhin. 2006. "Chieftaincy: An Overview." In *Chieftaincy in Ghana: Culture, Governance and Development*, edited by Irene K. Odotei and A. K. Awedoba, 27–42. Accra: Sub-Saharan Publishers.

Brenner, Neil. 2018. "Debating Planetary Urbanization: For an Engaged Pluralism." *Environment and Planning D: Society and Space* 36 (3): 570–590.

Brenner, Neil, and Nikos Katsikis. 2020. "Operational Landscapes: Hinterlands of the Capitalocene." *The Landscapists: Redefining Landscape Relations* 90 (1): 22–31.

Brigden, Kevin, Iryna Labunska, David Santillo, and Paul Johnston. 2008. *Chemical Contamination at E-Waste Recycling and Disposal Sites in Accra and Koforidua, Ghana*. Amsterdam, The Netherlands: Greenpeace International.

Brighenti, A. 2010. "On Territorology: Towards a General Science of Territory." *Culture and Society* 27 (1): 52–72.

Brooke, James. 1988. "Waste Dumpers Turning to West Africa." *New York Times*, July 17. http://www.nytimes.com/1988/07/17/world/waste-dumpers-turning-to-west-africa.html?pagewanted=all.

Brosius, Christiane. 2018. "Review of Mediating Mobility: Visual Anthropology in the Age of Migration, by Steffen Köhn." *Visual Anthropology Review* 34 (2): 166–168.

Broto, Vanesa Castán, and Martín Sanzana Calvet. 2020. "Sacrifice Zones and the Construction of Urban Energy Landscapes in Concepción, Chile." *Journal of Political Ecology* 27: 279–299.

Brown, Wendy. 2015. *Undoing the Demos: Neoliberalism's Stealth Revolution*. New York: Zone Books.

Browne, Simone. 2015. *Dark Matters: On the Surveillance of Blackness*. Durham, NC: Duke University Press.

Brulle, Robert. 2000. *Agency, Democracy, and Nature: The U.S. Environmental Movement from a Critical Theory Perspective*. Cambridge, MA: MIT Press.

Bullard, Robert. 1990. *Dumping in Dixie: Race, Class, and Environmental Quality*. Boulder, CO: Westview Press.

Burrell, Jenna. 2012. *Invisible Users: Youth in the Internet Cafés of Urban Ghana*. Cambridge, MA: MIT Press.

Burrell, Jenna. 2016. "What's the Real Story with Africa's E-Waste?" *Berkeley Blog, Opinion* (blog), September 1. https://news.berkeley.edu/berkeley_blog/whats-the-real-story-with-africas-e-waste/.

Burton, Andrew. 2005. *African Underclass: Urbanization, Crime, and Order in Dar es Salaam*. Athens: Ohio University Press.

Butchart, Alexander. 1998. *The Anatomy of Power: European Constructions of the African Body*. London and New York: Zed Books.

Butler, Judith, and Athena Athanasiou. 2013. *Dispossession: The Performative in the Political*. Cambridge: Polity.

Byster, Leslie A., and Ted Smith. 2006. "The Electronics Production Life Cycle: From Toxics to Sustainability: Getting Off the Toxic Treadmill." In *Challenging the Chip: Labor Rights and Environmental Justice in the Global Electronics Industry*, edited by Ted Smith, David A. Sonnenfeld, and David Naguib Pellow, 205–214. Philadelphia: Temple University Press.

Byster, Leslie A., and Wen-Ling Tu. 2006. "Electronic Waste and Extended Producer Responsibility." In *Challenging the Chip: Labor Rights and Environmental Justice in the Global Electronics Industry*, edited by Ted Smith, David A. Sonnenfeld, and David Naguib Pellow, 201–204. Philadelphia: Temple University Press.

Campbell, Colin. 1992. "The Desire for the New: Its Nature and Social Location as Presented in Theories of Fashion and Modern Consumerism." In *Consuming Technologies: Media and Information in Domestic Spaces*, edited by Roger Silverstone and Eric Hirsch, 53–54. London: Routledge.

Caravanos, Jack, Edith E. Clark, Richard Fuller, and Calah Lambertson. 2011. "Assessing Worker and Environmental Chemical Exposure Risks at an E-Waste Recycling and Disposal Site in Accra, Ghana." *Journal of Health and Pollution* 1: 16–25.

Caravanos, Jack, Edith E. Clarke, Carl S. Osei, and Yaw Amoyaw-Osei. 2013. "Exploratory Health Assessment of Chemical Exposures at E-Waste Recycling and Scrapyard Facility in Ghana." *Journal of Health and Pollution* 3: 11–22.

Caritas Ghana. 2018. "Caritas Ghana Kick Starts E-Waste Project." March 3. https://www.caritas-ghana.org/index.php/2018/03/03/caritas-ghana-kick-starts-e-waste-project/.

Carney, Liz. 2006. "Nigeria Fears E-Waste 'Toxic Legacy.'" BBC News, December 19. http://news.bbc.co.uk/2/hi/africa/6193625.stm.

Castells, M. 1997. *The Power of Identity*. Vol. 2, *The Information Age: Economy, Society, and Culture*. Oxford: Blackwell.

Cazabon, D., J. Fobil, G. Essegbey, and N. Basu. 2017. "Structured Identification of Response Options to Address Environmental Health Risks at the Agbogbloshie Electronic Waste Site." *Integrated Environmental Assessment and Management* 13 (6): 980–991.

Chakrabarty, Dipesh. 2019. *The Crises of Civilization: Explorations in Global and Planetary Histories*. New Delhi: Oxford University Press.

Chalfin, Brenda. 2010. *Neoliberal Frontiers: An Ethnography of Sovereignty in West Africa*. Chicago: University of Chicago Press.

Chalfin, Brenda. 2014. "Public Things, Excremental Politics, and the Infrastructure of Bare Life in Ghana's City of Tema." *American Ethnologist* 41 (1): 92–109.

Chalfin, Brenda. 2017. "'Wastelandia': Infrastructure and the Commonwealth of Waste in Urban Ghana." *Ethnos* 82 (4): 648–671.

Chalfin, Brenda. 2019. "On-Shore, Off-Shore Takoradi: Terraqueous Urbanism, Logistics, and Oil Governance in Ghana." *Environment and Panning D: Society and Space* 37 (5): 814–832.

Checker, Melissa. 2005. *Polluted Promises: Environmental Racism and the Search for Justice in a Southern Town*. New York: New York University Press.

Checker, Melissa. 2020. *The Sustainability Myth: Environmental Gentrification and the Politics of Justice*. New York: New York University Press.

Cho, Renee. 2016. "The Damaging Effects of Black Carbon." Columbia Climate School, Columbia University, March 22. https://news.climate.columbia.edu/2016/03/22/the-damaging-effects-of-black-carbon/.

Christophers, Brett. 2020. *Rentier Capitalism: Who Owns the Economy, and Who Pays for It?* London: Verso.

Clapp, Jennifer. 2001. *Toxic Exports: The Transfer of Hazardous Waste from Rich to Poor Countries*. Ithaca, NY: Cornell University Press.

Cliggett, Lisa. 2014. "Access, Alienation, and the Production of Chronic Liminality: Sixty Years of Frontier Settlement in a Zambian Park Buffer Zone." *Human Organization* 73 (2): 128–40.

Cock, Jacklyn. 2015. "How the Environmental Justice Movement Is Gathering Momentum in South Africa." *The Conversation*, November 1. https://

theconversation.com/how-the-environmental-justice-movement-is-gathering-momentum-in-south-africa-49819.

Collins, Timothy W. 2008. "The Political Ecology of Hazard Vulnerability: Marginalization, Facilitation and the Production of Differential Risk to Urban Wildfires in Arizona's White Mountains." *Journal of Political Ecology* 15: 21–43.

Comaroff, Jean, and John Comaroff. 1992. *Ethnography and the Historical Imagination.* Boulder, CO: Westview Press.

Comaroff, Jean, and John Comaroff. 2012. *Theory from the South: Or, How Euro-America Is Evolving Toward Africa.* New York: Routledge.

Comaroff, Jean, and John Comaroff. 2016. *The Truth About Crime: Sovereignty, Knowledge, Social Order.* Chicago: University of Chicago Press.

Comaroff, John L. 1997. "Images of Empire, Contests of Conscience: Models of Colonial Domination in South Africa." In *Tensions of Empire,* edited by Frederick Cooper and Ann Laura Stoler, 163–197. Berkeley: University of California Press.

Copper Development Association. 2020. "CDA Position Statement on Coronavirus (COVID-19) Pandemic." https://www.copper.org/applications/antimicrobial/COVID-19.html.

Cordner, Alissa. 2016. *Toxic Safety: Flame Retardants, Chemical Controversies, and Environmental Health.* New York: Columbia University Press.

Corvellec, Hervé, Steffen Böhm, Alison Stowell, and Francisco Valenzuela. 2020. "Introduction to the Special Issue on the Contested Realities of the Circular Economy." *Culture and Organization* 26 (2): 96–102.

Cresswell, T. 1996. *In Place, Out of Place: Geography, Ideology and Transgression.* Minneapolis, MN, and London: University of Minnesota Press.

Cresswell, T. 2010. "Towards a Politics of Mobility." *Environment and Planning D: Society and Space* 28 (1): 17–31.

Crocker, R., and K. Chiveralls. 2019. *Subverting Consumerism: Reuse in an Accelerated World.* London and New York: Routledge.

Cross, Jamie, and Declan Murray. 2018. "The Afterlives of Solar Power: Waste and Repair Off the Grid in Kenya." *Energy Research and Social Science* 44: 100–109.

Cullen, J. M. 2017. "Circular Economy: Theoretical Benchmark or Perpetual Motion Machine?" *Journal of Industrial Ecology* 21 (3): 483–486.

Dannenberg, R. O., J. M. Maurice, and G. M. Potter. 1973. *Recovery of Precious Metals from Electronic Scrap.* Report of Investigations 7683. Washington, DC: US Department of the Interior, Bureau of Mines.

Das, Veena, and Shalini Randeria. 2015. "Politics of the Urban Poor: Aesthetics, Ethics, Volatility, Precarity." Supplement, *Current Anthropology* 56 (S11): S3–S14.

Daum, K., J. Stoler, and R. J. Grant. 2017. "Toward a More Sustainable Trajectory for E-Waste Policy: A Review of a Decade of E-Waste Research in Accra, Ghana." *International Journal of Environmental Research and Public Health* 14 (135): 1–18.

Davidson, J., M. Smith, L. Bondi, and E. Probyn. 2008. "Emotion, Space, and Society: Editorial Introduction." *Emotion, Space and Society* 1 (1): 1–3.

Davies, Thom. 2013. "A Visual Geography of Chernobyl: Double Exposure." *International Labor and Working-Class History* 84: 116–139.

Davies, Thom. 2018. "Photography and Toxic Pollution: Exposing a Chemical Company." *Science as Culture* 27 (4): 543–551.

Davies, Thom. 2019. "Slow Violence and Toxic Geographies: 'Out of Sight' to Whom?" *Environment and Planning C: Politics and Space.* doi: 10.1177/2399654419841063.

Davis, John-Micheal, Grace Akese, and Yaakov Garb. 2019. "Beyond the Pollution Haven Hypothesis: Where and Why Do E-Waste Hubs Emerge and What Does This Mean for Policies and Interventions?" *Geoforum* 98: 36–45.

Davis, Mike. 2006. *Planet of Slums*. London and New York: Verso.

Davis, Mike. 2020. *The Monster Enters: COVID-19, Avian Flu and the Plagues of Capitalism*. New York and London: OR Books.

Davis, Mike, and Daniel Bertrand Monk, eds. 2008. *Evil Paradises: Dreamworlds of Neoliberalism*. New York and London: New Press.

Davis, Mike, and Jon Weiner. 2020. *Set the Night on Fire: L.A. in the 60s*. London: Verso.

Delaporte, Pablo Seward, and Sonia A. P. Grant. 2019. "Searches for Livability: An Interview with Adriana Petryna." Supplementals, *Fieldsights*, February 6. https://culanth.org/fieldsights/searches-for-livability-an-interview-with-adriana-petryna.

Desmond, Matthew. 2016. *Evicted: Poverty and Profit in the American City*. New York: Penguin.

Deutsche Gesellshaft fur Internationale Zusammenarbeit. 2019. *E-Waste Training Manual*. A Document of the German Cooperation and the Ghana Ministry of Environment, Science, Technology and Innovation. Bonn, Germany: Deutsche Gesellshaft fur Internationale Zusammenarbeit.

Dickson, K. B. 1968. "Background to the Problem of Economic Development in Northern Ghana." *Annals of the Association of American Geographers* 58 (4): 686–696.

Dickson, K. B. 1975. "Development Planning and National Integration in Ghana." In *The Search for National Integration in Africa*, edited by D. R. Smock and K. Bentsi-Enchill, 100–116. New York: Free Press.

Dietrich, Alexa S. 2014. *The Drug Company Next Door: Pollution, Jobs, and Community Health in Puerto Rico*. New York: New York University Press.

Dillon, Lindsey, Dawn Walker, Nicholas Shapiro, Vivian Underhill, Megan Martenyi, Sara Wylie, et al. 2017. "Environmental Data Justice and the Trump Administration: Reflections from the Environmental Data and Governance Initiative." *Environmental Justice* 10 (6): 186–192.

Dobler, Gregor, and Rita Kesselring. 2019. "Swiss Extractivism: Switzerland's Role in Zambia's Copper Sector." *Journal of Modern African Studies* 57 (2): 223–245.

Dolan, Catherine, and Dinah Rajak, eds. 2016. *The Anthropology of Corporate Social Responsibility*. New York and Oxford: Berghahn.

Donson, Tony. 2014. "Agbogbloshie: Digital Wasteland." *Dazed Magazine*, February 26. http://www.dazeddigital.com/photography/article/19008/1/agbogbloshie-digital-wasteland.

Droney, Damien. 2014. "Ironies of Laboratory Work During Ghana's Second Age of Optimism." *Cultural Anthropology* 29 (2): 363–384.

Droney, Damien. 2015. "Weedy Science: The Cultural Politics of Herbal Medicine Science in Ghana." PhD diss., Stanford University.

Du Bois, W. E. B. (1957) 2014. *The Black Flame Trilogy: Book One, The Ordeal of Mansart*. The Oxford W. E. B. Du Bois. Edited by Henry Louis Gates. Oxford: Oxford University Press.

Du Bois, W. E. B. (1959) 2014. *The Black Flame Trilogy: Book Two, Mansart Builds a School*. The Oxford W. E. B. Du Bois). Edited by Henry Louis Gates. Oxford: Oxford University Press.

Du Bois, W. E. B. (1961) 2014. *The Black Flame Trilogy: Book Three, Worlds of Color*. The Oxford W. E. B. Du Bois. Edited by Henry Louis Gates. Oxford: Oxford University Press.

Dürr, Eveline, and Rivke Jaffe. 2012. "Theorizing Slum Tourism: Performing, Negotiating and Transforming Inequality." *European Review of Latin American and Caribbean Studies* 93: 113–123.

Earle, Timothy. 1997. *How Chiefs Come to Power: The Political Economy in Prehistory.* Stanford, CA: Stanford University Press.

Earthworks and Oxfam America. 2004. *Dirty Metals: Mining, Communities and the Environment.* Washington, DC: Earthworks and Oxfam America.

Edwards, Paul. 2021. "Platforms Are Infrastructures on Fire." In Your Computer Is on Fire, edited by T. S. Mullaney, B. Peters, M. Hicks, and K. Philip, 313–336. Cambridge, MA: MIT Press.

Ellison, N. 1997. "Towards a New Social Politics: Citizenship and Reflexivity in Late Modernity." *Sociology* 31 (4): 697–717.

Ericksen, T. H., and E. Schober. 2017. "Waste and the Superfluous: An Introduction." *Social Anthropology* 25: 282–287.

Ernest, Tsifodze. 2020. "Slums, Coronavirus: The Remedy." *GhanaWeb*, April 14. https://www.ghanaweb.com/GhanaHomePage/features/Slums-coronavirus-The-Remedy-923401.

Escobar, Arturo. 2008. *Territories of Difference: Place, Movement, Life, Redes.* Durham, NC: Duke University Press.

Escobar, Arturo. 2018. *Designs for the Pluriverse: Radical Interdependence, Autonomy, and the Making of Worlds.* Durham, NC, and London: Duke University Press.

Faber, Daniel. 2008. *Capitalizing on Environmental Injustice.* New York: Rowman and Littlefield.

Fanon, Frantz. 1952. *Black Skin, White Masks.* New York: Grove Press.

Fanon, Frantz. 1963. *The Wretched of the Earth.* Translated by Richard Philcox. New York: Grove Press.

Farmer, Paul. 2003. *Pathologies of Power: Health, Human Rights, and the New War on the Poor.* Berkeley: University of California Press.

Farouk, Braimah R., and Mensah Owusu. 2012. "'If in Doubt, Count': The Role of Community-Driven Enumerations in Blocking Eviction in Old Fadama, Accra." *Environment and Urbanization* 24 (1): 47–57.

Farquhar, Judith, and Margaret Lock. 2007. "Introduction." In *Beyond the Body Proper: Reading the Anthropology of Material Life*, edited by M. Lock and J. Farquhar, 1–16. Durham, NC: Duke University Press.

Feldt, T., J. N. Fobil, and J. Wittseipe. 2014. "High Levels of PAH-Metabolites in Urine of E-Waste Recycling Workers from Agbogbloshie, Ghana." *Science of the Total Environment* 466–467: 369–376.

Ferguson, Gary. 2017. *Land on Fire: The New Reality of Wildfire in the West.* Portland, OR: Timber Press.

Ferguson, James. 1999. *Expectations of Modernity: Myths and Meanings of Urban Life on the Zambian Copperbelt.* Berkley: University of California Press.

Ferguson, James. 2005. "Seeing Like an Oil Company: Space, Security, and Global Capital in Neoliberal Africa." *American Anthropologist* 107 (3): 377–382.

Ferguson, James. 2006. *Global Shadows: Africa in the Neoliberal World Order.* Durham, NC, and London: Duke University Press.

Ferguson, James. 2010. "The Uses of Neoliberalism." *Antipode* 41: 166–184.

Ferguson, James. 2015. *Give a Man a Fish: Reflections on the New Politics of Distribution.* Durham, NC, and London: Duke University Press.

Fernandez-Stark, Karina, Penny Bamber, and Martin Walter. 2020. "COVID-19 and the New Age of Copper: Opportunities for Latin America." *VOX*, October 7. https://voxeu.org/article/covid-19-and-new-age-copper.

Fine, B., and K. Boateng. 2000. "Labour and Employment Under Structural Adjustment." In *Economic Reforms in Ghana: The Miracle and the Mirage*, edited by Ernest Aryeetey, Jane Harrigan, and Machiko Nissanke. Oxford: James Currey.

Fishbein, Betty. 2001. *Waste in the Wireless World*. New York: Inform.

Foucault, Michel. 1988. *The Technologies of the Self: A Seminar with Michel Foucault.* Amherst: University of Massachusetts Press.

Fowler, Cynthia. 2013. *Ignition Stories: Indigenous Fire Ecology in the Indo-Australian Monsoon Zone*. Durham, NC: Carolina Academic Press.

Fredericks, Rosalind. 2018. *Garbage Citizenship: Vital Infrastructures of Labor in Dakar, Senegal*. Durham, NC: Duke University Press.

Fuhriman, Darrell N. 2008. "Dangerous Donations: Discarded Electronics in Accra, Ghana." Master's thesis, Penn State University.

Gabrielson, Teena. 2019. "The Visual Politics of Environmental Justice." *Environmental Humanities* 11 (1): 27–51.

Gabrys, Jennifer. 2011. *Digital Rubbish: A Natural History of Electronics*. Ann Arbor: University of Michigan Press.

Gabrys, Jennifer. 2016. *Program Earth: Environmental Sensing Technology and the Making of a Computational Planet*. Minneapolis: University of Minnesota Press.

Gandy, Mathew. 2006. "Planning, Anti-planning and the Infrastructure Crisis Facing Metropolitan Lagos. *Urban Studies* 43 (2): 371–396.

Gareau, Brian J. 2012. "The Limited Influence of Global Civil Society." *Environmental Politics* 21 (1): 88–107.

Gareau, Brian J. 2013. *From Precaution to Profit: Contemporary Challenges to Environmental Protection in the Montreal Protocol*. New Haven, CT: Yale University Press.

Garrett, Laurie. 2020. "Trump Has Sabotaged America's Coronavirus Response." *Foreign Policy*, January 31.

Gavua, Kodzo, and Wazi Apoh. 2011. *Salvage Archaeology at the Bui Dam Project*. Document prepared for the Bui Power Authority. Legon: Department of Archaeology and Heritage Studies, University of Ghana.

Geschiere, Peter, and Josef Gugler. 1998. "The Urban–Rural Connection: Changing Issues of Belonging and Identification." *Africa* 68 (3): 309–319.

Getachew, Adom. 2019. *Worldmaking After Empire: The Rise and Fall of Self-Determination*. Princeton, NJ: Princeton University Press.

Geyer, Roland, Jenna R. Jambeck, and Kara Lavender Law. 2017. "Production, Use, and Fate of All Plastics Ever Made." *Science Advances* 3(7): e1700782.

Ghana Broadcasting Corporation. 2017. "Science in Africa." July 11.

Ghana Health Service. 2017. *2016 Annual Report*. Accra: Ghana Health Service.

Ghertner, D. Asher. 2020. "Airpocalypse: Distributions of Life Amidst Delhi's Polluted Airs." *Public Culture* 32 (1): 133–162.

Gieryn, T. 2000. "A Space for Place in Sociology." *Annual Review of Sociology* 26: 463–496.

Gille, Zsuzsa. 2007. *From the Cult of Waste to the Trash Heap of History: The Politics of Waste in Socialist and Postsocialist Hungary*. Bloomington: Indiana University Press.

Girdner, E. J., and J. Smith. 2002. *Killing Me Softly: Toxic Waste, Corporate Profit and the Struggle for Environmental Justice*. New York: Monthly Review Press.

Goldstein, Jesse. 2018. *Planetray Improvement: Cleantech Entrepreneurship and the Contradictions of Green Capitalism*. Cambridge, MA: MIT Press.

Gómez-Barris, Macarena. 2017. *The Extractive Zone: Social Ecologies and Decolonial Perspectives*. Durham, NC, and London: Duke University Press.

Gordillo, Gastón. 2014. *Rubble: The Afterlife of Destruction*. Durham, NC: Duke University Press.

Graeber, David. 2011. *Debt: The First 5,000 Years*. New York: Melville House.

Graeber, David, and Marshall Sahlins. 2017. *On Kings*. Chicago: HAU Books.

Grant, Richard. 2006. "Out of Place? Global Citizens in Local Spaces: A Study of the Informal Settlements in the Korle Lagoon Environs in Accra, Ghana." *Urban Forum* 17: 1–24.

Grant, Richard. 2016. "The 'Urban Mine' in Accra, Ghana." In *Out of Sight, Out of Mind: The Politics and Culture of Waste*, edited by Christof Mauch. Issue 2016/1 of *RCC Perspectives: Transformations in Environment and Society* 1: 21–29.

Grant, Richard, and Martin Oteng-Ababio. 2012. "Mapping the Invisible and Real 'African' Economy: Urban E-Waste Circuitry." *Urban Geography* 33: 1–21.

Grant, Richard, and Martin Oteng-Ababio. 2016. "The Global Transformation of Materials and the Emergence of Informal Urban Mining in Accra, Ghana." *Africa Today* 62 (4): 3–20.

Grant, Kristen, Fiona C. Goldizen, Peter D. Sly, Marie-Noel Brune, Maria Neira, Martin van den Berg, et al. 2013. "Health Consequences of Exposure to E-Waste: A Systematic Review." *Lancet Global Health* 1 (6): 350–361.

Green Advocacy Ghana. 2018. Facebook Post. January 22. https://www.facebook.com/Green-Advocacy-Ghana-158067427737532.

Green Advocacy Ghana. 2021. "GreenAd Ghana: About Us." Accessed June 23. https://greenadghana.com/about-us-1/.

Green Alert. 2010. Sakumono, Ghana: Green Advocacy Ghana and the Centre for Environment and Health, Research and Training.

Green, Rob G. 2019. "Ghana: A Burning Problem." *American Scholar*, September 3. https://theamericanscholar.org/ghana-a-burning-problem/.

Greengard, Samuel. 2015. *The Internet of Things*. Cambridge, MA: MIT Press.

Gregson, N., and M. Crang. 2010. "Materiality and Waste: Inorganic Vitality in a Networked World." *Environment and Planning A* 42: 1026–1032.

Grischow, Jeff. 2006. *Shaping Tradition: Civil Society, Community and Development in Colonial Northern Ghana, 1899–1957*. Leiden, The Netherlands: Brill.

Grossman, Elizabeth. 2006. *High Tech Trash: Digital Devices, Hidden Toxics, and Human Health*. Washington, DC: Island Press.

Gupta, Akhil. 2015. "An Anthropology of Electricity from the Global South." *Cultural Anthropology* 30 (4): 555–568.

Gupta, Akhil, and James Ferguson. 1997. "After 'Peoples and Cultures.'" In *Culture, Place, Power: Explorations in Critical Anthropology*, edited by A. Gupta and J. Ferguson, 1–29. Durham, NC: Duke University Press.

Gupta, Akhil, and Aradhana Sharma. 2006. "Globalization and Postcolonial States." *Current Anthropology* 47 (2): 277–307.

Gutberlet, Jutta, and Sebastián Carenzo. 2020. "Waste Pickers at the Heart of the Circular Economy: A Perspective of Inclusive Recycling from the Global South." *Worldwide Waste: Journal of Interdispclinary Studies* 3 (1): 1–14.

Gutberlet, Jutta, S. Carenzo, J. H. Kain, and A. Mantovani Martiniano de Azevedo. 2017. "Waste Picker Organizations and Their Contribution to the Circular Economy: Two Case Studies from a Global South Perspective." *Resources* 6 (4): 52.

Gyau-Boakye, P. 2001. "Environmental Impacts of the Akosombo Dam and Effects of Climate Change on the Lake Levels." *Environment, Development and Sustainability* 3 (1): 17–29.

Hannerz, Ulf. 1980. *Exploring the City: Inquiries Toward an Urban Anthropology.* New York: Columbia University Press.

Hannerz, Ulf. 1996. *Transnational Connections: Culture, People, Places.* New York: Routledge.

Haraway, Donna. 1988. "Situated Knowledges: The Science Question in Feminism and the Privilege of Partial Perspective." *Feminist Studies* 14 (3): 575–599.

Haraway, Donna. 2020. *Donna Haraway: Story Telling for Earthly Survival.* Documentary film, written and directed by Fabrizio Terranova. Brussels: Belgium Cinema.

Harden, Blaine. 2001. "The Dirt in the New Machine." *New York Times*, August 13.

Harding, Sandra. 1991. "Who Knows? Identities and Feminist Epistemology." In *(En) Gendering Knowledge: Feminists in Academe*, edited by Joan E. Hartman and Ellen Messer-Davidow, 100–115. Knoxville: University of Tennessee Press.

Hare, Nathan. 1970. "Black Ecology." *The Black Scholar* 1: 2–8.

Harms, Erik. 2013. "Eviction Time in the New Saigon: Temporalities of Displacement in the Rubble of Development." *Cultural Anthropology* 28 (2): 344–368.

Harper, Douglas. 2002. "Talking About Pictures: A Case for Photo Elicitation." *Visual Studies* 17: 13–16.

Harper, Krista. 2012. "Visual Interventions and the 'Crisis in Representation' in Environmental Anthropology: Researching Environmental Justice in a Hungarian Romani Neighborhood." *Human Organization* 71 (3): 292–305.

Harris, Lee. 2020. "Cuomo Pushes to Weaken Ban on Toxic Foam Burning." *New York Focus*, November 9.

Harrison, Graham. 2011. "Practices of Intervention: Repertoires, Habits, and Conduct in Neoliberal Africa." *Journal of Intervention and Statebuilding* 4 (4): 433–452.

Hart, Keith. 1978. "Informal Income Opportunities and Urban Employment in Ghana." *Journal of Modern African Studies* 11 (1): 61–89.

Harvey, David. 1973. *Social Justice and the City.* Baltimore: Johns Hopkins University Press.

Harvey, David. 2008. "The Right to the City." *New Left Review* 53: 23–40.

Harvey, David. 2009. *Social Justice and the City*, rev. ed. Athens: University of Georgia Press.

Harvey, David. 2012. *Rebel Cities: From the Right to the City to the Urban Revolution.* London: Verso.

Harvey, David. 2020. "Value in Motion." *New Left Review* 126. https://newleftreview.org/issues/ii126/articles/david-harvey-value-in-motion.

Harvey, Penelope. 2017. "Waste Futures: Infrastructures and Political Experimentation in Southern Peru." *Ethnos* 82 (4): 672–89.

Harvey, Penny. 2018. "Infrastructures in and out of Time: The Promise of Roads in Contemporary Peru." In *The Promise of Infrastructure*, edited by Nikhil Anand, Akhil Gupta, and Hannah Appel, 80–101. Durham, NC: Duke University Press.

Hastrup, Kirsten. 1995. *A Passage to Anthropology: Between Experience and Theory.* London and New York: Routledge.

Hawkins, Gay. 2006. *The Ethics of Waste: How We Relate to Rubbish*. Lanham, MD: Rowman and Littlefield.

Heacock, Michelle, Carol Bain Kelly, Kwadwo Ansong Asante, Linda S. Birnbaum, Åke Lennart Bergman, Marie-Noel Bruné, et al. 2016. "E-Waste and Harm to Vulnerable Populations: A Growing Problem." *Environmental Health Perspectives* 124 (5): 550–555.

Hearn, Julie. 1985. "The NGO-isation of Kenyan Society: USAID and the Restructuring of Health Care." *Review of African Political Economy* 25: 89–100.

Hearn, Julie. 2002. "The 'Invisible' NGO: US Evangelical Missions in Kenya." *Journal of Religion in Africa* 32 (1): 32–60.

Hecht, Gabrielle. 2018. "Interscalar Vehicles for and African Anthropocene: On Waste, Temporality, and Violence." *Cultural Anthropology* 33 (1): 109–141.

Hecht, Gabrielle. 2020. "Human Crap." *Aeon*, March 25. https://aeon.co/essays/the-idea-of-disposability-is-a-new-and-noxious-fiction.

Helmreich, Stefan. 1998. *Silicon Second Nature: Culturing Artificial Life in a Digital World*. Berkeley: University of California Press.

Hill, P. 1956. *The Gold Coast Cocoa Farmer*. London: Oxford University Press.

Hodgetts, Matthew, and Kevin McGravey. 2020. "Climate Change and the Free Marketplace of Ideas?" *Environmental Values* 29 (6): 713–752.

Hodgson, Dorothy L., and Judith A. Byfield. 2017. "Why Global Africa?" In *Global Africa: Into the Twenty First Century*, edited by Dorothy L. Hodgson and Judith A. Byfield, 1–9. Berkley: University of California Press.

Höges, Clemens. 2009. "The Children of Sodom and Gomorrah: How Europe's Discarded Computers Are Poisoning Africa's Kids." *Spiegel*, December 4.

Hogue, Cheryl. 2020. "Incineration May Spread, not Break Down PFAS. *Chemical and Engineering News*, April 27.

Holmes, Douglas R. 2000. *Integral Europe: Fast-Capitalism, Multiculturalism, Neofascism*. Princeton, NJ: Princeton University Press.

Holsey, Bayo. 2008. *Routes of Remembrance: Refashioning the Slave Trade in Ghana*. Chicago: University of Chicago Press.

Holsey, Bayo. 2013. "Black Atlantic Visions: History, Race, and Transnationalism in Ghana." *Cultural Anthropology* 28: 504–518.

Home, Andy. 2017. "What China's Imports Tell Us About the Copper Market." *Reuters*. September 28. https://www.reuters.com/article/china-copper-ahome/column-what-chinas-imports-tell-us-about-the-copper-market-andy-home-idINL8N1M94XH.

Hong, Sungmin, Jean-Pierre Candalone, Clair C. Patterson, and Claude F. Boutron. 1996. "History of Ancient Copper Smelting Pollution During Roman and Medieval Times Recorded in Greenland Ice." *Science* 272: 246–248.

Hoover, Elizabeth. 2017. *The River Is in Us: Fighting Toxics in a Mohawk Community*. Minneapolis: University of Minnesota Press.

Hoover, Elizabeth. 2019. "'Fires Were Lit Inside Them': The Pyropolitics of Water Protector Camps at Standing Rock." *Review of International American Studies* 12 (1): 11–43.

Horne, Gerald. 2020. *The Dawning of the Apocalypse: The Roots of Slavery, White Supremacy, Settler Colonialism, and Capitalism in the Long Sixteenth Century*. New York: Monthly Review Press.

Hosbey, Justin, Hilda Lloréns, and J. T. Roane, eds. Forthcoming. "Global Black Ecologies." *Environment and Society: Advances in Research*.

Hosoda, Junki, John Ofosu-Anim, Edward Benjamin Sabi, Lailah Gifty Akita, Siaw Onwona-Agyeman, Rei Yamashita, et al. 2014. "Monitoring of Organic Micropollutants in Ghana by Combination of Pellet Watch with Sediment Analysis: E-Waste as a Source of PCBs." *Marine Pollution Bulletin* 86: 575–581.

Howe, Cymene, and Dominic Boyer. 2020. "Verdant Optimism: On How Capitalism Will Never Save the World." Theorizing the Contemporary, *Fieldsites*, March 24. https://culanth.org/fieldsights/verdant-optimism-on-how-capitalism-will-never-save-the-world.

Hsu, Hsuan L. 2020. *The Smell of Risk: Environmental Disparities and Olfactory Aesthetics.* New York: New York University Press.

Hsu, Wendy F. 2014. "Digital Ethnography Toward Augmented Empiricism: A New Methodological Framework." *Journal of Digital Humanities* 3 (1): 1–5.

Huang, Jingyu, Philip Nti Nkrumah, Desmond Ofosu Anim, and Ebenezer Mensah. 2014. "E-Waste Disposal Effects on the Aquatic Environment: Accra, Ghana." *Reviews of Environmental Contamination and Toxicology* 229: 19–34.

Hughes, David McDermott. 2013. "Climate Change and the Victim Slot: From Oil to Innocence." *American Anthropologist* 115 (4): 570–581.

Hugo, P. 2010. "A Global Graveyard for Dead Computers in Ghana" (photo essay). *New York Times Magazine*, August 4.

Hutnyk, John. 2004. "Photogenic Poverty: Souvenirs and Infantilism." *Journal of Visual Culture* 3 (1): 77–94.

Huws, Ursula. 1999. "Material World: The Myth of the 'Weightless Economy.'" *Socialist Register 1999: Global Capitalism vs. Democracy*, edited by Leo Panitch and Colin Leys, 35:29–55. London: Merlin Press.

Ingold, Tim. 2017. *Anthropology in/as Education.* London and New York: Routledge.

Interagency Task Force on Electronic Stewardship. 2014. *Moving Sustainable Electronics Forward: An Update to the National Strategy for Electronics Stewardship.* Washington, DC: White House Council on Environmental Quality, Environmental Protection Agency, and General Services Administration.

International Labour Organization. 2019. "Decent Work in the Management of Electrical and Electronic Waste (E-Waste). Issues paper for the Global Dialogue Forum on Decent Work in the Management of Electrical and Electronic Waste (E-Waste) (Geneva, 9–11 April 2019). https://www.ilo.org/sector/Resources/publications/WCMS_673662/lang--en/index.htm.

Isenhour, Cindy, and Joshua Reno. 2019. "On Materiality and Meaning: Ethnographic Engagements with Reuse, Repair and Care." *Worldwide Waste: Journal of Interdisciplinary Studies* 2 (1): 1–8.

Ivancheva, Mariya, and Kathryn Keating. 2020. "Revisiting Precarity, with Care: Productive and Reproductive Labour in the Era of Flexible Capitalism." *Ephemera: Theory and Politics in Organization* 20 (4): 251–282.

Jacka, Jerry. 2018. "The Anthropology of Mining: The Social and Environmental Impacts of Resource Extraction in the Mineral Age." *Annual Review of Anthropology* 47: 66–77.

Jaffe, Rivke. 2016. *Concrete Jungles: Urban Pollution and the Politics of Difference in the Caribbean.* New York: Oxford University Press.

Jalbert, Kirk, Anna Willow, David Casagrande, and Stephanie Paladino, eds. 2017. *ExtrACTION: Impacts, Engagements, and Alternative Futures.* New York: Routledge.

Jaramillo, Pablo. 2020. "Mining Leftovers: Making Futures on the Margins of Capitalism." *Cultural Anthropology* 35 (1): 48–73.

Jasanoff, Sheila. 2015. "Future Imperfect: Science, Technology, and the Imaginations of Modernity." In *Dreamscapes of Modernity: Sociotechincal Imaginaries and the Fabrication of Power*, edited by Sheila Jasanoff and Sang-Hyun Kim, 1–33. Chicago: University of Chicago Press.

Jasanoff, Sheila, and Sang-Hyun Kim, eds. 2015. *Dreamscapes of Modernity: Sociotechincal Imaginaries and the Fabrication of Power*. Chicago: University of Chicago Press.

Jobson, Ryan Cecil. 2019. "The Case for Letting Anthropology Burn: Sociocultural Anthropology in 2019." *American Anthropologist* 122 (2): 259–271.

Jones, Tim. 2016. *The Fall and Rise of Ghana's Debt: How a New Debt Trap Has Been Set*. Accra: Integrated Social Development Centre Ghana; London: Jubilee Debt Campaign UK; Accra: SEND Ghana; Accra: VAZOBA Ghana, All-Afrikan Networking Community Link for International Development; Peki, Ghana: Kilombo Centre for Citizens' Rights and Conflict Resolution; Accra; Abibimman Foundation Ghana.

Kane, A. 2002. "Senegal's Village Diaspora and the People Left Ahead." In *The Transnational Family*, edited by Deborah Bryceson and Ulla Vuorela, 245–264. New York: Bloomsbury.

Kaplan, Jeremy. 2014. "Welcome to Hell: Photographer Documents Africa's E-Waste Nightmare." Fox News, March 6. https://www.foxnews.com/tech/welcome-to-hell-photographer-documents-africas-e-waste-nightmare.

Kara, Siddharth. 2018. "Is Your Phone Tainted by the Misery of the 35,000 Children in Congo's Mines?" *The Guardian*, October 12.

Karkari, A. Y., K. A. Asante, and C. A. Biney. 2006. "Water Quality Characteristics at the Estuary of Korle Lagoon in Ghana." *West African Journal of Applied Ecology* 10 (1): 1–12.

Kaza, Silpa, Lisa Yao, Perinaz Bhada-Tata, and Frank Van Woerden. 2018. *What a Waste 2.0: A Global Snapshot of Solid Waste Management to 2050*. Urban Development Series. Washington, DC: World Bank.

Kean, Fergal. 1998. "Another Picture of Starving Africa: It Could Have Been Taken in 1984, or 1998." *The Guardian*, June 8.

Keirsten, S., and P. Michael. 1999. *A Report on Poison PCs and Toxic TVs*. San Jose, CA: Silicon Valley Toxics Coalition.

Kelley, Robin D. 2002. *Freedom Dreams: The Radical Black Imagination*. Boston: Beacon Press.

Kilanski, Kristine, and Javier Auyero. "Introduction." In *Violence at the Urban Margins*, edited by Javier Auyero, Phillipe Bourgois, and Nancy Scheper-Hughes, 1–20. Oxford: Oxford University Press.

Kimble, David. 1963. *A Political History of Ghana: The Rise of Gold Coast Nationalism, 1850–1928*. Oxford: Clarendon Press.

Kimura, Aya H., and Abby Kinchy. 2019. *Science by the People: Participation, Power, and the Politics of Environmental Knowledge*. New Brunswick, NJ: Rutgers University Press.

Kirby, Jon P. 2003. "Peacebuilding in Northern Ghana: Cultural Themes and Ethnic Conflict." In *Ghana's North: Research on Culture, Religion, and Politics of Societies in Transition*, edited by F. Kroger and B. Meier, 161–205. Frankfurt: Peter Lang.

Kirby, Peter Wynne, and Anna Lora-Wainwright. 2015. "Peering Through Loopholes, Tracing Conversions: Remapping Transborder Trade in Electronic Waste." *Area* 47 (1): 4–6.

Kirsch, Stuart. 2014. *Mining Capitalism: The Relationship Between Corporations and Their Critics*. Berkeley: University of California Press.

Kleespies, E. K., J. P. Bennetts, and T. A. Henrie. 1970. "Gold Recovery from Scrap Electronic Solders by Fused-Salt Electrolysis." *Journal of Metals* 22 (1): 42–44.

Klein, Naomi. 2019. *On Fire: The (Burning) Case for a Green New Deal*. New York: Simon and Schuster.

Klein, Naomi. 2020a. "How Big Tech Plans to Profit from the Pandemic. *The Guardian*, May 13.

Klein, Naomi. 2020b. "Coronavirus Capitalism—And How to Beat It." *The Intercept*, March 16.

Klein, P. 2009. "Ghana: Digital Dumping Ground." *Frontline*, episode aired June 23. https://www.pbs.org/frontlineworld/stories/ghana804/.

Kleinman, Arthur. 1995. *Writing at the Margin: Discourse Between Anthropology and Medicine*. Berkeley: University of California Press.

Kleinman, Arthur, Veena Das, and Margaret Lock. 1997. "Introduction." In *Social Suffering*, edited by Arthur Kleinman, Veena Das, and Margaret Lock, ix–xxvii. Berkeley: University of California Press.

Klooster, Wim, and Alfred Padula. 2005. *The Atlantic World: Essays on Slavery, Migration, and Imagination*. New Jersey: Pearson/Prentice Hall.

Kloster, S., N. M. Mahowald, J. T. Randerson, and P. J. Lawrence. 2012. "The Impacts of Climate, Land Use, and Demography on Fires During the 21st Century Simulated by CLM-CN." *Biogeosciences* 9: 509–525.

Knapp, F. 2016. "The Birth of the Flexible Mine: Changing Geographies of Mining and the E-Waste Commodity Frontier." *Environment and Planning A* 48 (10): 1889–1909.

Kokutse, Francis. 2020. "Ghana's Main Opposition Party Will Contest Election Results." *Washington Post*, December 10.

Komarova, Milena, and Maruška Svašek, eds. 2018. *Ethnographies of Movement, Sociality and Space*. New York and Oxford: Berghan.

Konadu, Kwasi, and Clifford C. Campbell. 2016a. "Introduction." In *The Ghana Reader: History, Culture, Politics*, edited by Kwasi Konadu and Clifford C. Campbell, 1–16. Durham, NC: Duke University Press.

Konadu, Kwasi, and Clifford C. Campbell. 2016b. "Between the Sea and the Savanna, 1500–1700." In *The Ghana Reader: History, Culture, Politics*, edited by Kwasi Konadu and Clifford C. Campbell, 81–82. Durham, NC: Duke University Press.

Koné, Lassana. 2009. "Pollution in Africa: A New Toxic Waste Colonialism? An Assessment of Compliance of the Bamako Convention in Côte d'Ivoire." LLM diss., University of Pretoria.

Konings, Martijn. 2015. *The Emotional Logics of Capitalism: What Progressives Have Missed*. Stanford, CA: Stanford University Press.

Koser, K., ed. 2003. *New African Diasporas*. London: Routledge.

Kyere, Vincent Nartey, Klaus Greve, and Sampson M. Atiemo. 2016. "Spatial Assessment of Soil Contamination by Heavy Metals from Informal Electronic Waste Recycling in Agbogbloshie, Ghana." *Environmental Health and Toxicology* 31: e2016006.

Labban, Mazan. 2014. "Deterritorializing Extraction: Bioaccumulation and the Planetary Mine." *Annals of the Association of American Geographers* 104 (3): 560–576.

Lamoreaux, J. 2016. "What if the Environment Is a Person? Lineages of Epigenetic Science in a Toxic China." *Cultural Anthropology* 31 (2): 188–214.

Larkin, Brian. 2018. "Promising Forms: The Political Aesthetics of Infrastructure." In *The Promise of Infrastructure*, edited by Nikhil Anand, Akhil Gupta, and Hannah Appel, 175–202. Durham, NC: Duke University Press.

Larmer, Brook. 2018. "E-Waste Offers an Economic Opportunity as Well as Toxicity." *New York Times Magazine*, July 5. Accessed July 6, 2018. https://www.nytimes.com/2018/07/05/magazine/e-waste-offers-an-economic-opportunity-as-well-as-toxicity.html.

Lartey, Nii Larte. 2018. "Agbogbloshie Demolition: AMA 'Bullying' Residents to Increase Revenue." August 26. https://citinewsroom.com/2018/08/agbogbloshie-demolition-ama-bullying-residents-to-increase-revenue-nii-lante/.

Lashaw, Amanda, Christian Vannier, and Steven Sampson, eds. 2017. *Cultures of Doing Good: Anthropologists and NGOs*. Tuscaloosa: University of Alabama Press.

Laskaris, Zoey, Chad Milando, Stuart Batterman, Bhramar Mukherjee, Niladri Basu, Marie S. O'Neill, et al. 2019. "Derivation of Time-Activity Data Using Wearable Cameras and Measures of Personal Inhalation Exposure Among Workers at an Informal Electronic-Waste Recovery Site in Ghana." *Annals of Work Exposures and Health* 63 (8): 829–841.

Latour, Bruno. 1993. *We Have Never Been Modern*. Translated by C. Porter. Cambridge, MA: Harvard University Press.

Lawal, Shola. 2019. "Nigeria Has Become an E-Waste Dumpsite for Europe, US, and Asia." TRT World, February 15. Accessed June 2, 2019. https://www.trtworld.com/magazine/nigeria-has-become-an-e-waste-dumpsite-for-europe-us-and-asia-24197.

Leach, Melissa, Ian Scoones, and Brian Wynne. 2005. "Introduction: Science, Citizenship, and Globalization." In *Science and Citizens: Globalization and the Challenge of Engagement*, edited by Melissa Leach, Ian Scoones, and Brian Wynne, 3–14. London: Zed Books.

Lefebvre, Henri. 1970. *La Révolution Urbaine*. Paris: Gallimard.

Lefebvre, Henri. 1991. *The Production of Space*. Oxford: Oxford University Press.

Lefebvre, Henri. 2003. *The Urban Revolution*. Translated by Roberto Bononno. Minneapolis: University of Minnesota Press.

Lepawsky, Josh. 2012. "Legal Geographies of E-Waste Legislation in Canada and the US: Jurisdiction, Responsibility and the Taboo of Production." *Geoforum* 43 (6): 1194–1206.

Lepawsky, Josh. 2014. "'E-Waste': Mapping a Controversy." *The Rubbish Bin*. Reassembling Rubbish. May 6. https://scalar.usc.edu/works/reassembling-rubbish/e-waste-mapping-a-controversy?path=the-rubbish-bin.

Lepawsky, Josh. 2018. *Reassembling Rubbish: Worlding Electronic Waste*. Cambridge, MA: MIT Press.

Lepawsky, Josh. 2019. "PSA: Beware of Easy Narratives." *Discard Studies* (blog), August 12. https://discardstudies.com/2019/08/12/psa-beware-of-easy-narratives/.

Lepawsky, Josh, and Grace Akese. 2015. "Sweeping away Agbogbloshie Again." *Discard Studies* (blog), June 23. Accessed January 5, 2018. https://discardstudies.com/2015/06/23/sweeping-away-agbogbloshie-again/.

Lepawsky, Josh, and Mostaem Billah. 2011. "Making Chains That (Un)Make Things: Waste-Value Relations and the Bangladeshi Rubbish Electronics Industry." *Geografiska Annaler* 93 (2): 121–139.

Lepawsky, J., and C. McNabb. 2010. "Mapping International Flows of Electronic Waste." *Canadian Geographer* 54 (2): 177–195.

Lepawsky, Josh. n.d. Entry in Discard Studies Compendium. https://discardstudies.com/discard-studies-compendium/.

Lepawsky, Josh. n.d. Reassembling Rubbish Website. https://scalar.usc.edu/works/reassembling-rubbish/index.

Leve, Lauren, and Lamia Karim. 2001. "Introduction: Privatizing the State: Ethnography of Development, Transnational Capital, and NGOs." *Political and Legal Anthropology Review* 24 (1): 53–58.

Lewis, Barbara. 2019. "Uganda's Tungsten Mines Sue International Tin Association for Defamation." *Reuters*, May 28.

Liboiron, Max. 2013. "The Atemporality of 'Ruin Porn': The Carcass and the Ghost by Sarah Wanenchak." *Discard Studies* (blog), July 17. https://discardstudies.com/2013/07/17/the-atemporality-of-ruin-porn-the-carcass-the-ghost-by-sarah-wanenchak/.

Liboiron, Max. 2014. "Myopic Spatial Politics in Dominant Narratives of E-Waste." *Discard Studies* (blog), March 6. https://discardstudies.com/2014/03/06/myopic-spatial-politics-in-dominant-narratives-of-e-waste/.

Liboiron, Max. 2015. "The Perils of Ruin Porn: Slow Violence and the Ethics of Representation." *Discard Studies* (blog), March 23. https://discardstudies.com/2015/03/23/the-perils-of-ruin-porn-slow-violence-and-the-ethics-of-representation/.

Liboiron, Max. 2016. "Redefining Pollution and Action: The Matter of Plastics." *Journal of Material Culture* 21 (1): 87–110.

Liboiron, Max. 2018a. "The What and the Why of Discard Studies." *Discard Studies* (blog), September 1. https://discardstudies.com/2018/09/01/the-what-and-the-why-of-discard-studies/.

Liboiron, Max. 2018b. "Waste Colonialism." *Discard Studies* (blog), November 1. https://discardstudies.com/2018/11/01/waste-colonialism/.

Liboiron, Max. 2021. *Pollution Is Colonialism*. Durham, NC: Duke University Press.

Linder, C., and M. Meissner, eds. 2016. *Global Garbage: Urban Imaginaries of Waste, Excess, and Abandonment*. New York: Routledge.

Lipovetsky, G. 2011. "The Hyper-Consumption Society." In *Beyond the Consumption Bubble*, edited by K. M. Ekström and B. Glans, 25–36. London: Routledge.

Little, Peter C. 2009. "Negotiating Community Engagement and Science in the Federal Environmental Public Health Sector." *Medical Anthropology Quarterly* 23 (2): 94–118.

Little, Peter C. 2012. "Another Angle on Pollution Experience: Toward an Anthropology of the Emotional Ecology of Risk Mitigation." *Ethos* 40 (4): 431–452.

Little, Peter C. 2013. "Vapor Intrusion: The Political Ecology of an Emerging Environmental Health Concern." *Human Organization* 72 (2): 121–131.

Little, Peter C. 2014. *Toxic Town: IBM, Pollution, and Industrial Risks*. New York: New York University Press.

Little, Peter C. 2015. "Toxic Struggle and Corporate Paradox in a High-Tech Industrial Birthplace." *Toxic News* (blog), November 4. https://toxicnews.org/2015/11/04/toxic-struggle-and-corporate-paradox-in-a-high-tech-industrial-birthplace/.

Little, Peter C. 2016. "On Electronic Pyropolitics and Pure Earth Friction in Agbogbloshie." *Toxic News* (blog), November 8. https://toxicnews.org/2016/11/08/on-electronic-pyropolitics-and-pure-earth-friction-in-agbogbloshie/.

Little, Peter C. 2017. "On the Micropolitics and Edges of Survival in a Technocapital Sacrifice Zone." *Capitalism, Nature, Socialism* 4: 62–77.

Little, Peter C. 2019. "Bodies, Toxins, and E-Waste Labour Interventions in Ghana: Toward a Toxic Postcolonial Corporality?" *Revista de Antropología Iberoamericana* 14 (1): 51–71.

Little, Peter C. 2020. "Witnessing E-Waste Through Participatory Photography in Ghana." In *Environmental Justice and Citizen Science in a Post-Truth Age*, edited by Thom Davies and Alice Mah, 140–157. Manchester, UK: Manchester University Press.

Little, Peter C., and Grace A. Akese. 2019. "Centering the Korle Lagoon: Exploring Blue Political Ecologies of E-Waste in Ghana." *Journal of Political Ecology* 26 (1): 448–465.

Little, Peter C., and Cristina Lucier. 2017. "Global Electronic Waste, Third Party Certification Standards, and Resisting the Undoing of Environmental Justice Politics." *Human Organization* 76 (3): 204–214.

Liu, Junxiao, Xijin Xu, Kusheng Wu, Zhongxian Piao, Jinrong Huang, Yongyong Guo, et al. 2011. "Association Between Lead Exposure from Electronic Waste Recycling and Child Temperament Alterations." *Neurotoxicology* 32: 458–464.

Lock, Margaret. 2015. "Comprehending the Body in the Era of the Epigenome." *Current Anthropology* 56 (2): 151–177.

Lock, Margaret, and Judith Farquhar, eds. 2007. *Beyond the Body Proper: Reading the Anthropology of Material Life*. Durham, NC: Duke University Press.

Locke, David. 1990. *Drum Damba: Talking Drum Lessons*. Tempe, AZ: White Cliffs Media.

Lonely Planet. n.d. "Accra Old Fadama Slum Small-Group Walking Tour." https://www.lonelyplanet.com/ghana/accra/activities/accra-old-fadama-slum-small-group-walking-tour/a/pa-act/v-6501GHUB/355309.

Lora-Wainwright, A. 2017. "E-Waste Work: Hierarchies of Value and the Normalization of Pollution in Guiyu." In *Resigned Activism: Living with Pollution in Rural China*, edited by A. Lora-Wainwright, 125–156. Cambridge, MA: MIT Press.

Low, S., and D. Lawrence-Zúñiga. 2003. "Locating Culture." In *The Anthropology of Space and Place: Locating Culture*, edited by S. Low and D. Lawrence-Zúñiga, 1–48. Oxford: Blackwell.

Lundgren, Karin. 2012. *The Global Impact of E-Waste: Addressing the Challenge*. Geneva: International Labour Organization.

Lutz, Catherine, and Jane Collins. 1993. *Reading National Geographic*. Chicago: University of Chicago Press.

Lyons, Santiago. 2017. "The Purpose of Photography in a Post-Truth Era." *Time*, January 26. Accessed February 8, 2018. http://time.com/4650956/photojournalism-post-truth/.

MacBride, S. 2011. *Recycling Reconsidered: The Present Failures and Future Promise of Environmental Action in the United States*. Cambridge, MA: MIT Press.

MacGaffey, J., and R. Bazanguissa-Ganga. 2000. *Congo–Paris: Transnational Traders on the Margins of the Law*. London: International African Institute.

MacGaffey, Wyatt. 2006. "A History of Tamale, 1907–1957 and Beyond." *Transactions of the Historical Society of Ghana*, New Series (10): 109–124.

Mamdani, Mahmood. 1996. *Citizen and Subject: Contemporary Africa and the Legacy of Late Colonialism*. Princeton, NJ: Princeton University Press.

Marcus, George E., and Michael M. J. Fischer. 1986. *Anthropology as Cultural Critique: An Experimental Moment in the Human Sciences*. Chicago: University of Chicago Press.

Marder, Michael. 2015. *Pyropolitics: When the World Is Ablaze*. London and Lanham, MD: Rowman and Littlefield.

Masco, Joseph. 2016. "The Crisis in Crisis." Supplement, *Current Anthropology* 58 (S15): S65–S76.

Masuda J. R., B. Poland, and J. Baxter. 2010. "Reaching for Environmental Health Justice: Canadian Experiences for a Comprehensive Research, Policy and Advocacy Agenda in Health Promotion." *Health Promotion International* 25 (4): 453–463.

Mbembe, Achille. 2001. *On the Postcolony*. Berkeley: University of California Press.

Mbembe, Achille. 2017. *Critique of Black Reason*. Translated by Laurent Dubois. Durham, NC: Duke University Press.

Mbembe, Achille. 2019. "Technology and Eschatology in the Computational Age." Public Lecture, Institute for Critical Social Inquiry, June 11. https://www.facebook.com/events/d41d8cd9/achille-mbembe-technology-and-eschatology-in-the-computational-age/1225886964236175/.

Mbembe, Achille. 2020a. "Interview by Malka Gouzer. Achille Mbembe: 'Ignorance too, is a form of power.'" Accessed November 8, 2020. https://www.chilperic.ch/interview/achille-mbembe-15.html.

Mbembe, Achille. 2020b. *Brutalisme*. Paris: La Découverte.

Mbembe, Achille. 2021. *Out of the Dark Night: Essays on Decolonization*. New York: Columbia University Press.

McAdam, Doug, Sidney Tarrow, and Charles Tilly. 2001. *Dynamics of Contention*. Cambridge: Cambridge University Press.

McGranahan, Carole. 2016. "Theorizing Refusal: An Introduction." *Cultural Anthropology* 31 (3): 319–325.

McNeil, J. R. 2000. *Something New Under the Sun: An Environmental History of the Twentieth-Century World*. New York and London: Norton.

Menon-Sen, K., and G. Bhan. 2008. *Swept off the Map: Surviving Eviction and Resettlement in Delhi*. New Delhi: Yoda Press.

Mignolo, Walter D., and Catherine Walsh. 2018. *On Decoloniality: Concepts, Analytics, Praxis*. Durham, NC: Duke University Press.

Millar, Kathleen M. 2018. *Reclaiming the Discarded: Life and Labor on Rio's Garbage Dump*. Durham, NC, and London: Duke University Press.

Miller, DeMond Shondell, and Nyjeer Wesley. 2016. "Toxic Disasters, Biopolitics, and Corrosive Communities: Guiding Principles in the Quest for Healing in Flint, Michigan." *Environmental Justice* 9 (3): 69–75.

Millington, Nate, and Mary Lawhon. 2019. "Geographies of Waste: Conceptual Vectors from the Global South." *Progress in Human Geography* 43 (6): 1044–1063.

Minor, Jesse, and Geoffrey A. Boyce. 2018. "Smokey Bear and the Pyropolitics of United States Forest Governance." *Political Geography* 62: 79–93.

Minorities at Risk Project. 2004. *Chronology for Mossi-Dagomba in Ghana*. Accessed January 10, 2019. https://www.refworld.org/docid/469f388f1e.html.

Minter, Adam. 2016. "The Burning Truth Behind an E-Waste Dump in Africa." *Smithsonian*, January 13. Accessed January 31, 2018. http://www.smithsonianmag.com/sciencenature/burning-truth-behind-e-waste-dump-africa-180957597.

Mitchell, Don. 2003. *The Right to the City: Social Justice and the Fight for Public Space*. New York: Guilford Press.

Mitchell, Timothy. 1989. "The World as Exhibition." *Comparative Studies in Society and History* 31 (2): 217–236.

Mitchell, Timothy. 2011. *Carbon Democracy: Political Power in the Age of Oil*. London: Verso.

Mitchell, William J. 2005. *Placing Words: Symbols, Space, and the City*. Cambridge, MA: MIT Press.

Mkhwanazi, Nolwazi. 2016. "Medical Anthropology in Africa: The Trouble with a Single Story." *Medical Anthropology* 35 (2): 193–202.

Mohai, Paul, and Bunyan Bryant. 1992. "Environmental Injustice: Weighing Race and Class as Factors in the Distribution of Environmental Hazards." *University of Colorado Law Review* 63 (4): 921–932.

Mongabay. 2012. "High-Tech Hell: New Documentary Brings Africa's E-Waste Slum to Life." *Mongabay Environmental News*, April 30,

Moore, Donald S., Anand Pandian, and Jake Kosek. 2003. "The Cultural Politics of Race and Nature: Terrains of Power and Practice." In *Race, Nature, and the Politics of Difference*, edited by D. S. Moore, J. Kosek, and A. Pandian, 1–70. Durham, NC, and London: Duke University Press.

Moore, Jason W., ed. 2016. *Anthropocene or Capitalocene? Nature, History, and the Crisis of Capitalism*. Oakland, CA: PM Press.

Moore, Jason W. 2019. "Making Sense of the Planetary Inferno: Planetary Justice in the Web of Life." Lecture at Garage Museum of Contemporary Art, Moscow, Russia, August 20. https://www.youtube.com/watch?v=AKxezfGWLsA.

Moore, Sarah A. 2011. "Global Garbage: Waste, Trash Trading, and Local Garbage Politics." In *Global Political Ecology*, edited by Richard Peet, Paul Robbins, and Michael J. Watts, 133–144. London: Routledge.

Moreno-Tejada, Jaime. 2020. "Editorial Introduction to the Special Collection 'Dirty Places, Geographies of Waste.'" *Worldwide Waste: Journal of Interdiscplinary Studies* 3 (1): 1–3.

Moritz, Max A., Marc-André Parisien, Enric Batllori, Meg A. Krawchuk, Jeff Van Dorn, David J. Ganz, et al. 2012. "Climate Change and Disruptions to Global Fire Activity." *Ecosphere* 3 (6): 49.

Morrison, Nicky. 2017. "Struggling for the Right to Be Recognized: The Informal Settlement of Old Fadama, Accra, Ghana." In *Geographies of Forced Eviction*, edited by K. Brickell, M. Fernández Arrigoitia, and A. Vasudevan, 25–45. London: Palgrave Macmillan.

Morton, Timothy. 2013. *Hyperobjects: Philosophy and Ecology After the End of the World*. Minneapolis and London: University of Minnesota Press.

Msila, Vuyisile, ed. 2020. *Developing Teaching and Learning in Africa: Decolonising Perspectives*. Stellenbosch, South Africa: African Sun Media.

Mullaney, Thomas S., Benjamin Peters, Mar Hicks, and Kavita Philip, eds. 2021. *Your Computer Is on Fire*. Cambridge, MA: MIT Press.

Muller, Jerry Z. 2018. *The Tyranny of Metrics*. Princeton, NJ: Princeton University Press.

Müller, Simone M. 2019. "Hidden Externalities: The Globalization of Hazardous Waste." *Business History Review* 93 (1): 51–74.

Mulvaney, Dustin. 2014. "Are Green Jobs Just Jobs? Cadmium Narratives in the Life Cycle of Photovoltaics." *Geoforum* 54: 178–186.

Mulvaney, Dustin. 2019. *Solar Power: Innovation, Sustainability, and Environmental Justice*. Berkeley: University of California Press.

Murphy, Michelle. 2017. *The Economization of Life*. Durham, NC, and London: Duke University Press.

Murphy, Michelle. 2018. "Against Population, Towards Afterlife." In *Making Kin not Population*, edited by Adele Clark and Donna Haraway, 101–124. Chicago: Prickly Paradigm Press.

Murphy, Michelle. 2020. "Some Keywords Towards Decolonial Methods: Studying Settler Colonial Histories and Environmental Violence from Tkaronto." *History and Theory* 59 (3): 376–384.

Mwai, Peter. 2021. "COVID-19 Africa: Who Is Getting the Vaccine?" BBC News, March 22. https://www.bbc.com/news/56100076.

Myers, Garth. 2011. *African Cities: Alternative Visions of Urban Theory and Practice.* London and New York: Zed Books.

Nading, Alex M. 2020. "Living in a Toxic World." *Annual Review of Anthropology* 49: 209–224.

Nagle, Robin. 2013. *Picking Up: On the Streets and Behind the Trucks with the Sanitation Workers of New York City.* New York: Farrar, Straus, and Giroux.

Ndlovu-Gatsheni, Sabelo. 2014. "What Is Beyond Discourses of Alterity? Reflections on the Constitution of the Present and the Construction of African Subjectivity." In *The Social Contract in Africa*, edited by Sanya Osha, 111–130. Bramfontein: Africa Institute of South Africa.

Neale, Timothy, and Jennifer Mairi Macdonald. 2019. "Permits to Burn: Weeds, Slow Violence, and the Extractive Future of Northern Australia." *Australian Geographer* 50 (4): 1–17.

Neale, Timothy, Alex Zahara, and Will Smith. 2019. "An Eternal Flame: The Elemental Governance of Wildfire's Pasts, Presents and Futures." *Cultural Studies Review* 25 (2): 115–134.

Newell, P., and D. Mulvaney. 2013. "The Political Economy of the Just Transition." *Geographical Journal* 179 (2): 132–134.

Nguyen, Vinh-Kim. 2009. "Government by Exception: Enrolment and Experimentality in Mass HIV Treatment Programs in Africa." *Social Theory and Health* 7 (3): 196–218.

Niewöhner, J. 2011. "Epigenetics: Embedded Bodies and the Molecularisation of Biography and Milieu." *BioSocieties* 6 (3): 279–298.

Nixon, Rob. 2011. *Slow Violence and the Environmentalism of the Poor.* Cambridge, MA: Harvard University Press.

Nkrumah, Kwame. 1965. *Neo-Colonialism: The Last Stage of Imperialism.* London: Nelson and Sons.

Nnorom, I. C., and O. Osibanjo. 2008. "Electronic Waste (E-Waste): Material Flows and Management Practices in Nigeria." *Waste Management* 28 (8): 1472–1479.

Nordstrom, Carolyn. 2007. *Global Outlaws: Crime, Money, and Power in the Contemporary World.* Berkeley: University of California Press.

Nriagu, J. O. 1990. "Global Metal Pollution." *Environment* 32 (7): 7–11, 28–33.

Nriagu, J. O. 1994. "Industrial Activity and Metal Emissions." In *Industrial Ecology and Global Change*, edited by R. Socolow, C. Andrews, F. Berkhout, and V. Thomas, 277–285. Cambridge: Cambridge University Press.

Nriagu, J. O. 1996. "A History of Global Metal Pollution." *Science* 272 (5259): 223–234.

Ntiamoa-Baidu, Y. 1991. "Seasonal Changes in the Importance of Coastal Wetlands in Ghana for Wading Birds." *Biological Conservation* 57: 139–158.

O'Donnell, Caroline, and Dillon Pranger, eds. 2020. *The Architecture of Waste: Design for a Circular Economy.* London: Routledge.

O'Keefe, Phil. 1988. "Toxic Terrorism." *Review of African Political Economy* 15 (42): 84–90.

Okereke, Chukwumerije. 2008. "Equity Norms in Global Environmental Governance." *Global Environmental Politics* 8 (3): 25–50.

Olivier de Sarda, Jean-Pierre. 1995. "A Moral Economy of Corruption in Africa." *Journal of Modern African Studies* 37 (1): 25–52.

Olson, Valerie. 2012. "Political Ecology in the Extreme: Asteroid Activism and the Making of an Environmental Solar System." *Anthropological Quarterly* 85 (4): 1027–1044.

Onuoha, Debbie. 2016. "Economies of Waste: Rethinking Waste Along the Korle Lagoon." *Journal for Undergraduate Ethnography* 6 (1): 1–16.

Oppong, Christine. 1973. *Growing Up in Dagbon*. Accra: Ghana.

Ortner, Sherry B. 1995. "Resistance and the Problem of Ethnographic Refusal." *Comparative Studies in Society and History* 37 (1): 173–193.

Osseo-Asare, DK. n.d. "Agbogbloshie Makerspace Platform." Accessed October 3, 2020. https://qamp.net/about/.

Osseo-Asare, DK. 2016. Quoted in "Assessing Agbogbloshies." *E-Scrap News*. July 12. https://resource-recycling.com/e-scrap/2016/07/12/assessing-agbogbloshie/.

Oteng-Ababio, M. 2010. "E-Waste: An Emerging Challenge for Solid Waste Management in Ghana." *International Development Planning Review* 32 (2): 191–206.

Oteng-Ababio, M. 2012. "When Necessity Begets Ingenuity: E-Waste Scavenging as a Livelihood Strategy in Accra, Ghana." *African Studies Quarterly* 13 (1/2): 1–21.

Oteng-Ababio, Martin, Ebenezer Forkuo Amankwaa, and Mary Anti Chama. 2014. "The Local Contours of Scavenging for E-Waste and Higher-Valued Constituent Parts in Accra, Ghana." *Habitat International* 43: 163–171.

Oteng-Ababio, M., and R. Grant. 2019. "Ideological Traces in Ghana's Urban Plans: How Do Traces Get Worked Out in the Agbogbloshie, Accra?" *Habitat International* 83: 1–10.

Ottaviani, Jacopo. 2018. "E-Waste Republic." European Journalism Centre document. Bill and Melinda Gates Foundation. https://interactive.aljazeera.com/aje/2015/ewaste/ https://www.bbc.com/news/56100076.

Owusu, M. 2013. "Community-Managed Reconstruction After the 2012 Fire in Old Fadama, Ghana." *Environment and Urbanization* 25: 243–248.

Oxfeld, Ellen, and Lynellyn D. Long. 2004. "Introduction: An Ethnography of Return." In *Coming Home? Refugees, Migrants, and Those Who Stayed Behind*, edited by Lynellyn D. Long and Ellen Oxfeld, 1–15. Philadelphia: University of Pennsylvania Press.

Panitch, Leo, and Greg Albo, eds. 2020. *Beyond Digital Capitalism: New Ways of Living*. Socialist Register 2021. London: Merlin Press.

Parajuly, Keshav, Ruediger Kuehr, Abhishek Kumar Awasthi, Colin Fitzpatrick, Josh Lepawsky, Elisabeth Smith, et al. 2019. *Future E-Waste Scenarios*. Bonn, Germany: StEP and United Nations University UNU-ViE SCYCLE; Osaka, Japan: United Nations Environment Programme IETC.

Parenti, Christian. 2011. *Tropic of Chaos: Climate Change and the New Geography of Violence*. New York: Nation Books.

Pearson, Thomas W. 2017. *When the Hills Are Gone: Frac Sand Mining and the Struggle for Community*. Minneapolis: University of Minnesota Press.

Pedersen, Isabel, and Andrew Iliadis. 2020. *Embodied Computing: Wearables, Implantables, Embeddables, Ingestibles*. Cambridge, MA: MIT Press.

Pellow, David N. 2001. "Environmental Justice and the Political Process: Movements, Corporations, and the State." *Sociological Quarterly* 42 (1): 47–67.

Pellow, David N. 2007. *Resisting Global Toxics: Transnational Movements for Environmental Justice*. Cambridge, MA: MIT Press.

Pellow, David N. 2016. "Toward a Critical Environmental Justice Studies: Black Lives Matter as an Environmental Justice Challenge." *Du Bois Review: Social Science Research on Race* 13 (2): 1–16.

Pellow, Deborah. 2008. *Landlords and Lodgers: Socio-Spatial Organization in an Accra Zongo*. Chicago: University of Chicago Press.

Pellow, Deborah. 2011. "Internal Transmigrants: A Dagomba Diaspora." *American Ethnologist* 38 (1): 132–147.

Pellow, Deborah. 2016. "Logics of Violence Among the Dagomba in Northern Ghana." In *The Management of Chieftaincy and Ethnic Conflicts in Ghana: Complementary Pathways and Competing Institutions*, edited by Steve Tonah and A. S. Anamzoya, 47–63. Accra: Woeli Pubs.

Peluso, Nancy Lee, and Michael Watts, eds. 2001. *Violent Environments*. Ithaca, NY: Cornell University Press.

Pérez, Rafael Fernández-Font. 2014. "Tools for Informal E-Waste Recyclers in Agbogbloshie, Ghana." Master's thesis, Royal Holloway, University of London.

Perry, Mark. 2014. "A Fire in the Mind of Arabs: The Arab Spring in Revolutionary History." *Insight Turkey* 16 (1): 27–34.

Petryna, Adriana. 2018. "Wildfires at the Edges of Science: Horizoning Work amid Runaway Change." *Cultural Anthropology* 33 (4): 570–595.

Picard, M. H., and T. Beigi. 2020. "Regimes of Waste (Im)Perceptibility in the Life Cycle of Metal." *Transnational Legal Theory* 11 (1–2): 197–218.

Pickren, Graham. 2014. "Political Ecologies of Electronic Waste: Uncertainty and Legitimacy in the Governance of E-Waste Geographies." *Environment and Planning A* 46 (1): 26–45.

Pierce, Joseph, Mary Lawhon, and Tyler McReary. 2020. "From Precarious Work to Obsolete Labour? Implications of Technological Disemployment for Geographical Scholarship." *Geografiska Annaler: Series B, Human Geography* 101 (2): 84–101.

Pierre, Jemina. 2013. *The Predicament of Blackness: Postcolonial Ghana and the Politics of Race*. Chicago: University of Chicago Press.

Pierre-Louis, Kendra. 2019. "Complex Fires Gain Ferocity as Earth Heats Up." *New York Times*. August 29. https://www.nytimes.com/2019/08/28/climate/fire-amazon-africa-siberia-worldwide.html.

Pink, Sarah, ed. 2007. *Visual Interventions*. New York: Berghahn Books.

Pink, Sarah, John Postill, Larissa Hjorth, Jo Tacchi, Tania Lewis, and Heather A. Horst. 2016. *Digital Ethnography: Principles and Practice*. Los Angeles: SAGE.

Pitt, Jeremy. 2020. "The Big Tech–Academia–Parliamentary Complex and Techno-Fuedalism." *IEEE Technology and Society Magazine*, September 24.

Plange, Nii-K. 1979. "Underdevelopment in Northern Ghana: Natural Causes or Colonial Capitalism?" *Review of African Political Economy* 6: 4–14.

Poisson, Chelsey, Sheri Boucher, Domenique Selby, Sylvia P. Ross, Charulata Jindal, Jimmy T. Efird, et al. 2020. "A Pilot Study of Airborne Hazards and Other Toxic Exposures in Iraq War Veterans." *International Journal of Environmental Research and Public Health* 17 (9): 1–15.

Poole, Deborah. 1997. *Vision, Race, and Modernity: A Visual Economy of the Andean Image World*. Princeton, NJ: Princeton University Press.

Povinelli, Elizabeth. 2018. *Geontologies: A Requiem to Late Liberalism*. Durham, NC: Duke University Press.

Povinelli, Elizabeth. 2019. "The Urban Intentions of Geontopower." *E-Flux Architecture*, May 3.

Powell, J. 2019. "Scientists Reach 100% Consensus on Anthropogenic Global Warming." *Bulletin of Science, Technology, and Society* 37 (4): 183–184.

Power, M. J., J. Marlon, N. Ortiz, P. J. Bartlein, S. P. Harrison, F. E. Mayle, et al. 2008. "Changes in Fire Regimes Since the Last Glacial Maximum: An Assessment Based on a Global Synthesis and Analysis of Charcoal Data." *Climate Dynamics* 30: 887–907.

Prakash, Siddharth, and Andreas Manhart. 2010. *Socio-economic Assessment and Feasibility Study on Sustainable E-Waste Management in Ghana*. Freiburg, Germany: Öko-Institut.

Pratt, Mary Louise. 1992. *Imperial Eyes: Travel Writing and Transculturation*. London: Routledge.

Precarity Lab. 2020. *Technoprecarious*. London: Goldsmiths Press.

Prince, Ruth J. 2014. "Introduction: Situating Health and the Public in Africa." In *Making and Unmaking Public Health in Africa: Ethnographic and Historical Perspectives*, edited by Ruth J. Prince and Rebecca Marshland, 1–54. Athens: Ohio University Press.

Prince, Ruth J., and Rebecca Marshland, eds. 2014. *Making and Unmaking Public Health in Africa: Ethnographic and Historical Perspectives*. Athens: Ohio University Press.

Prüss-Ustün, A., J. Wolf, C. Corvalán, R. Bos, and M. Neira. 2016. *Preventing Disease Through Healthy Environments: A Global Assessment of the Burden of Disease from Environmental Risks*. Geneva: World Health Organization.

Pure Earth. 2014. "Change and Hope Comes to Agbogbloshie." Press release, October 22. https://www.pureearth.org/change-hope-comes-agbogbloshie/.

Pure Earth. 2015. "Ghana (Agbogbloshie)—E-Waste Recycling." Project update. Accessed September 5, 2017. https://www.pureearth.org/project/agbobloshie-e-waste/.

Pyne, Stephene. 1982. *Fire in America: A Cultural History of Wildland and Rural Fire*. Princeton: Princeton University Press.

Pyne, Stephen. 1994. "Maintaining Focus: An Introduction to Anthropogenic Fire." *Chemosphere* 29 (5): 889–911.

Pyne, Stephen. 1995. *World Fire: The Culture of Fire on Earth*. Seattle: University of Washington Press.

Pyne, Stephen. 2001. *Fire: A Brief History*. Seattle: University of Washington Press.

Pyne, Stephen. 2012. *Fire: Nature and Culture*. London: Reaktion Books.

Rams, Dagna. 2018. "Creating a Unity Across Space: Living Between Subsistence Agriculture and Scrap Work in Ghana." Unpublished manuscript.

Raphael, Chad, and Ted Smith. 2006. "Importing Extended Producer Responsibility for Electronic Equipment into the United States." In *Challenging the Chip: Labor Rights and Environmental Justice in the Global Electronics Industry*, edited by Ted Smith, David A. Sonnenfeld, and David Naguib Pellow, 247–259. Philadelphia: Temple University Press.

Rasmussen, Carol. 2015. "In Africa, More Smoke Leads to Less Rain." *NASA Earth Science News*, August, 5.

Rathbone, Richard. 2000. *Nkrumah and the Chiefs: The Politics of Chieftaincy in Ghana 1951–1960*. Athens: Ohio University Press.

Reddy, R. N. 2016. "Reimagining E-Waste Circuits: Calculation, Mobile Policies, and the Move to Urban Mining in Global South Cities." *Urban Geography* 37 (1): 57–76.

Reich, Robert. 2020. "Fire, Pestilence and a Country at War with Itself: The Trump Presidency Is Over." *The Guardian*, May 31.

Renfrew, Daniel. 2018. *Life Without Lead: Contamination, Crisis, and Hope in Uruguay.* Berkeley: University of California Press.

Reno, Joshua. 2011. "Beyond Risk: Emplacement and the Production of Environmental Evidence." *American Ethnologist* 38 (3): 516–530.

Reno, Joshua. 2014. "Toward a New Theory of Waste: From 'Matter out of Place' to Signs of Life." *Theory, Culture, and Society* 31 (6): 3–27.

Reno, Joshua. 2015. "Waste and Waste Management." *Annual Review of Anthropology* 44: 557–572.

Reno, Joshua. 2016. *Waste Away: Working and Living with a North American Landfill.* Berkeley: University of California Press.

Repetto, Robert. 2004. *Silence Is Golden, Leaden, and Copper: Disclosure of Material Environmental Information in the Hard Rock Mining Industry.* Report 1. New Haven, CT: Yale School of Forestry and Environmental Studies.

Riederer, Anne M., Stephanie Adrian, and Ruediger Kuehr. 2013. "Assessing the Health Effects of Informal E-Waste Processing." *Journal of Health and Pollution* 4: 1–3.

Ro, Christine. 2019. "The World's Most Toxic Eggs." *Forbes*, May 28.

Roane, J. T., and Justin Hosbey. 2019. "Mapping Black Ecologies." *Current Research in Digital History* 2. https://crdh.rrchnm.org/essays/v02-05-mapping-black-ecologies/.

Robbins, Joel. 2013. "Beyond the Suffering Subject: Toward an Anthropology of the Good." *Journal of the Royal Anthropological Institute* 19 (3): 447–462.

Roberson, Ed. 2021. *Asked What Has Changed.* Middleton, CT: Wesleyan University Press.

Roberts, Jody A., and Nancy Langston, eds. 2008. "Toxic Bodies/Toxic Environments: An Interdisciplinary Forum." *Environmental History* 13 (4): 629–703.

Robinson, Cedric J. 1983. *Black Marxism: The Making of the Black Radical Tradition.* London: University of North Carolina Press. Rodgers, Dennis, and Bruce O'Neil. 2012. "Infrastructural Violence: Introduction to the Special Issue." Special issue, *Ethnography* 13 (4): 401–412.

Rosen, Lauren Coyle. 2020. *Fires of Gold: Law, Spirit, and Sacrificial Labor in Ghana.* Berkeley: University of California Press.

Rosenfeld, Heather, Sarah Moore, Eric Nost, Robert E. Roth, and Kristen Vincent. 2018. "Hazardous Aesthetics: A 'Merely Interesting' Toxic Tour of Waste Management Data." *GeoHumanities* 4 (1): 262–281.

Rottenburg, Richard. 2009. "Social and Public Experiments and New Figurations of Science and Politics in Postcolonial Africa." *Postcolonial Studies* 12 (4): 423–440.

Saethre, Eirik. 2020. *Wastelands: Recycled Commodities and the Perpetual Displacement of Ashkali and Romani Scavengers.* Berkeley: University of California Press.

Santos, Bonaventure de Souza. 2014. *Epistemologies of the South: Justice Against Epistemicide.* New York, Routledge.

Satariano, Adam. 2019. "How the Internet Travels Across Oceans." *New York Times*, March 10.

Scheper-Hughes, Nancy. 1995. "The Primacy of the Ethical: Propositions for a Militant Anthropology." *Current Anthropology* 36 (3): 409–440.

Schiller, Ben. 2013. "A Look Inside the Hellscape of One of the World's Largest Electronic Waste Dumps." *Fast Company*, December 3.

Schipper, Branco W., Hsiu-Chuan Lin, Marco A. Meloni, Kjell Wansleeben, Reinout Heijungs, and Ester van der Voet. 2018. "Estimating Global Copper Demand Until 2100 with Regression and Stock Dynamics." *Resource, Conservation and Recycling* 132: 28–36.

Schluep, Mathias, Tatiana Terekhova, Andreas Manhart, Esther Muller, David Rochat, and Oladele Osibanjo. 2012. "Where Are WEEE in Africa? Electronics Gone Green." Paper presented at the 2012+ Conference, Berlin, Germany, September 9–12.

Schor, J. B. 1998. *The Overspent American: Upscaling, Downshifting, and the New Consumer*. New York: Basic Books.

Schuller, Mark. 2009. "Gluing Globalization: NGOs as Intermediaries in Haiti." *Political and Legal Anthropology Review* 32 (1): 84–104.

Schulz, Yvan. 2015. "Towards a New Waste Regime? Critical Reflections on China's Shifting Market for High-Tech Discards." *China Perspectives* 3: 43–50.

Schulz, Yvan. 2018. "Modern Waste: The Political Ecology of E-Scrap Recycling in China." PhD diss., University of Neuchâtel.

Schulz, Y., and A. Lora-Wainwright. 2019. "In the Name of Circularity: Environmental Improvement and Business Slowdown in a Chinese Recycling Hub." *Worldwide Waste: Journal of Interdisciplinary Studies* 2 (1): 1–13.

Schwab, Tim. 2020. "Are Bill Gate's Billions Distorting Public Health Data?" *The Nation*, December 3.

Shiva, Vandana. 2020. "Towards a Transformation of Hope for the Earth." Interview by Stefania Romano. DiEM25 TV. May 19. https://www.youtube.com/watch?v=uIePOnP4eGk.

Simpson, A. 2007. "On Ethnographic Refusal: Indigeneity, 'Voice', and Colonial Citizenship." *Junctures* 9: 67–80.

Simpson, A. 2014. *Mohawk Interruptus: Political Life Across the Borders of Settler States*. Durham, NC: Duke University Press.

Sinervo, Aviva, and Michael D. Hill. 2011. "The Visual Economy of Andean Childhood Poverty." *Journal of Latin American and Caribbean Anthropology* 16 (1): 114–142.

Singer, Merrill. 2016. "Introduction." In *A Companion to the Anthropology of Environmental Health*, edited by Merrill Singer, 1–18. Chichester, UK: Wiley/Blackwell.

Slade, Giles. 2006. *Made to Break: Technology and Obsolescence in America*. Cambridge, MA: Harvard University Press.

Smith, James H. 2017. "What's in Your Cell Phone?" In *Global Africa: Into the Twenty-First Century*, edited by Dorothy Hodgson and Judith Byfield, 289–297. Berkeley: University of California Press.

Smith, Linda Tuhiwai. 2008. *Decolonizing Methodologies: Research and Indigenous Peoples*. London and New York: Zed Books.

Smith, Ted, David A. Sonnenfeld, and David Naguib Pellow, eds. 2006. *Challenging the Chip: Labor Rights and Environmental Justice in the Global Electronics Industry*. Philadelphia: Temple University Press.

Smith, Tony. 2000. *Technology and Capital in the Age of Lean Production: A Marxian Critique of the "New Economy."* New York: SUNY Press.

Songsore, J. 2009. *Regional Development in Ghana: The Theory and the Reality*. Accra: Woeli Publishers.

Srnicek, Nick. 2016. *Platform Capitalism*. Chichester, UK: Wiley.

Stacey, Paul. 2015. "Political Structure and the Limits of Recognition and Representation in Ghana." *Development and Change* 46 (1): 25–47.

Stacey, Paul. 2019. *State of Slum: Precarity and Informal Governance at the Margins in Accra*. London: Zed Books.

Stacey, Paul, and Christian Lund. 2016. "In a State of Slum: Governance in an Informal Urban Settlement in Ghana." *Journal of Modern African Studies* 54 (4): 591–615.

Stahl, George. 2021. "Ship Stuck in Suez Canal and Chip Shortages: What Global Supply-Chain Problems Mean for You." *Wall Street Journal*, March 26.

Staniland, Martin. 1975. *The Lions of Dagbon: Political Change in Northern Ghana*. Cambridge: Cambridge University Press.

Steffen, Will, Johan Rockström, Katherine Richardson, Timothy M. Lenten, Carl Folke, Diana Liverman, et al. 2018. "Trajectories of the Earth System in the Anthropocene." *Proceedings of the National Academy of Sciences of the United States of America* 115 (33): 8252–8259.

Stewart, Sheelagh. 1997. "Happy Ever After in the Marketplace: Non-government Organizations and Uncivil Society." *Review of African Political Economy* 24 (71): 11–34.

Stocker, Thomas F., Dahe Qin, Gian-Kasper Plattner, Melinda M. B. Tignor, Simon K. Allen, Judith Boschung, et al. 2013. *Climate Change 2013: The Physical Science Basis*. Contribution of Working Group I to the *Fifth Assessment Report of the Intergovernmental Panel on Climate Change*. Summary for Policymakers. Cambridge: Cambridge University Press.

Stone, Livia K. 2015. "Suffering Bodies and Scenes of Confrontation: The Art and Politics of Representing Structural Violence." *Visual Anthropology Review* 31 (2): 177–189.

Strathern, Marilyn. 2020. *Relations: An Anthropological Account*. Durham, NC: Duke University Press.

Suarez-Villa, Luis. 2009. *Technocapitalism: A Critical Perspective on Technological Innovation and Corporatism*. Philadelphia: Temple University Press.

Sustainable Recycling Industries. 2018. "Ghana's Way Towards Sustainable E-Waste Recycling." March 5. https://www.sustainable-recycling.org/ghanas-way-towards-sustainable-e-waste-recycling-first-country-in-africa-to-officially-launch-guidelines-for-environmentally-sound-e-waste-management/.

Swanson, Ana. 2019. "U.S. Delays Some China Tariffs Until Stores Stock Up for Holidays." *New York Times*, August 13.

Szasz, Andrew. 1994. *EcoPopulism: Toxic Waste and the Movement for Environmental Justice*. Minneapolis: University of Minnesota Press.

Sze, Julie. 2020. *Environmental Justice in a Moment of Danger*. Berkeley: University of California Press.

Tahiru, Abdul-Gafaru. 2019. "The Global and Regional Impact of Climate Change and Migration on Food Security and Development: A Case Study of Northern Ghana." PhD diss., Howard University.

Talton, Benjamin. 2016. "That All Konkomba Should Unite." In *The Ghana Reader: History, Culture, Politics*, edited by Kwasi Konadu and Clifford C. Campbell, 344–346. Durham, NC: Duke University Press.

Taylor, Chloe. 2020. " 'Welcome to the Age of Copper': Why the Coronavirus Pandemic Could Spark a Red Metal Rally." CNBC, June 24. https://www.cnbc.com/2020/06/24/coronavirus-why-the-pandemic-could-spark-a-copper-rally.html.

Thompson, Krista A. 2015. *Shine: The Visual Economy of Light in African Diasporic Aesthetic Practice*. Durham, NC: Duke University Press.

Thunberg, Greta. 2019. " 'Our House Is on Fire': Greta Thunberg, 16, Urges Leaders to Act on Climate." *The Guardian*, January 25.

Tichenor, Marlee. 2016. "The Power of Data: Global Malaria Governance and the Senegalese Data Retention Strike." In *Metrics: What Counts in Global Health*, edited by Vincanne Adams, 105–124. Durham, NC: Duke University Press.

Tojo, Naoko. 2006. "Design Change in Electrical and Electronic Equipment: Impacts of Extended Producer Responsibility Legislation in Sweden and Japan." In *Challenging the Chip: Labor Rights and Environmental Justice in the Global Electronics Industry*, Ted Smith, David A. Sonnenfeld, and David Naguib Pellow, 273–284. Philadelphia: Temple University Press.

Tosca, M. G., D. J. Diner, M. J. Garay, and O. V. Kalashnikova. 2015. "Human-Caused Fires Limit Convection in Tropical Africa: First Temporal Observations and Attribution." *Geophysical Research Letters* 45 (15): 6492–6501.

Tousignant, Noémi. 2018. *Edges of Exposure: Toxicology and the Problem of Capacity in Postcolonial Senegal*. Durham, NC: Duke University Press.

Townsend, Janet G., Gina Porter, and Emma Mawdsley. 2004. "Creating Spaces of Resistance: Development NGOs and Their Clients in Ghana, India, and Mexico." *Antipode* 36 (5): 871–889.

Trouillot, Michel-Rolph. 1982. "Motion in the System: Coffee, Color, and Slavery in Eighteenth-Century Saint-Domingue." *Review (Fernand Braudel Center)* 5 (3): 331–388.

Trouillot, Michel-Rolph. 1991. "Anthropology and the Savage Slot: The Poetics and Politics of Otherness." In *Recapturing Anthropology*, edited by Richard G. Fox, 17–44. Santa Fe, NM: School of American Research Press.

Tsikata, Dzodzi, and Wayo Seini. 2004. "Identities, Inequalities and Conflicts in Ghana." RISE Working Paper 5. Centre for Research on Inequality, Human Security and Ethnicity, Queen Elizabeth House, University of Oxford, Oxford.

Tsing, Anna. 2000. "Inside the Economy of Appearances." *Public Culture* 12 (1): 115–144.

Tsing, Anna Lowenhaupt. 2005. *Friction: An Ethnography of Global Connection*. Princeton, NJ: Princeton University Press.

Tsing, Anna. 2009. "Supply Chains and the Human Condition." *Rethinking Marxism: A Journal of Economics, Culture and Society* 21 (2): 148–176.

Tsing, A. 2014. "Blasted Landscapes (and the Gentle Arts of Mushroom Picking)." In *The Multispecies Salon*, edited by E. Kirksey, 87–110. Durham, NC: Duke University Press.

Tsing, Anna Lowenhaupt. 2015. *The Mushroom at the End of the World: On the Possibility of Life in Capitalist Ruin*. Princeton, NJ: Princeton University Press.

Tsing, Anna Lowenhaupt, Heather Anne Swanson, Elaine Gan, and Nils Bubandt, eds. 2017. *Arts of Living on a Damaged Planet*. Minneapolis: University of Minnesota Press.

United Nations. 2019. *World Investment Report, 2019*. Geneva: United Nations.

United Nations Communications Group. 2017. *The Sustainable Development Goals (SDGs) in Ghana: Why They Matter and How We Can Help*. New York: United Nations Communications Group.

United Nations Environment Programme. 2012. *21 Issues for the 21st Century: Result of the UNEP Foresight Process on Emerging Environmental Issues*. Nairobi, Kenya: United Nations Environment Programme.

United Nations Environment Programme. 2015. *Global Waste Management Outlook*. Vienna, Austria: International Solid Waste Association.

United Nations E-Waste Coalition. 2019. *A New Circular Vision for Electronics: Time for a Global Reboot*. Geneva: World Economic Forum.

Ureta, Sebastian. 2021. "Ruination Science: Producing Knowledge from a Toxic World." *Science, Technology, and Human Values* 46 (1): 29–52.

US International Trade Commission. 2013. *Used Electronic Products: An Examination of U.S. Exports.* Investigation No. 332–528. USITC Publication 4379. Washington, DC: US International Trade Commission.

van der Velden, Maja. 2019. "Zombie Statistics and Poverty Porn." Africa Is a Country. Accessed May 5, 2020 https://africasacountry.com/2019/04/zombie-statistics-and-poverty-porn.

van der Velden, Maja, and Martin Oteng-Ababio. 2019. Six Myths About Electronic Waste in Agbogbloshie, Ghana." Africa Is a Country. Accessed May 5, 2020. https://africasacountry.com/2019/03/six-myths-about-electronic-waste-in-agbogbloshie-ghana.

Van Dijk, R. 2003. "Religion, Reciprocity and Restructuring Family Responsibility in the Ghanaian Pentecostal Diaspora." In *The Transnational Family*, edited by Deborah Bryceson and Ulla Vuorela, 173–196. New York: Bloomsbury.

Varoufakis, Yanis. 2021. "Yanis Varoufakis: Capitalism Has Become 'Techno-Fuedalism.'" *Algazeera*, February 19.

Velis, C. 2017. "Waste Pickers in the Global South: Informal Recycling Sector in a Circular Economy Era. *Waste Management and Research* 35: 329–331.

Vergés, Françoise. 2017. "Racial Capitalocene: Is the Anthropocene Racial?" *Verso* (blog). August 30. https://www.versobooks.com/blogs/3376-racial-capitalocene.

Voyles, Traci Brynne. 2015. *Wastelanding: Legacies of Uranium Mining in Navajo Country.* Minneapolis: University of Minnesota Press.

Walkover, Lily. 2016. "When Good Works Count." In *Metrics: What Counts in Global Health*, edited by Vincanne Adams, 163–177. Durham, NC, and London: Duke University Press.

Walsh, Catherine E. 2020. "Decolonial Learnings, Askings, and Musings." *Postcolonial Studies* 23 (4): 604–611.

Wang, Jackie. 2018. *Carceral Capitalism*. South Pasadena, CA: Semiotext(e).

West, Cornell, and Christa Buschendorf. 2014. *Black Prophetic Fire*. Boston: Beacon Press.

White, Damian. 2020. "Just Transitions/Designs for Transitions: Preliminary Notes on a Design Politics for a Green New Deal." *Capitalism Nature Socialism* 31 (2): 20–39.

Wittsiepe, J., J. N. Fobil, and H. Till. 2015. "Levels of Polychlorinated Dibenzo-p-Dioxins, Dibenzofurans (PCDD/Fs) and Biphenyls (PCBs) in Blood of Informal E-Waste Recycling Workers from Agbogbloshie, Ghana, and Controls." *Environment International* 79: 65–73.

World Bank. 2003. *Project Performance Assessment Report: Ghana Mining Sector Rehabilitation Project (Credit 1921-GH) and Mining Sector Development and Environment Project (Credit 2743-GH).* Report 26197. Washington, DC: World Bank.

World Bank. 2019. *The World Bank Group Action Plan on Climate Change Adaptation and Resilience.* Washington, DC: World Bank.

World Health Organization. 2011. *Burn Prevention. Success Stories, Lessons Learned.* Geneva: World Health Organization.

Yaro, Joseph Awetori, Samuel Nii Ardey Codjoe, Samuel Adgei-Mensah, Akosua Darkwah, and Steven Owusu Kwankye. 2011. "Migration and Population Dynamics: Changing Community Formations in Ghana." Migration Studies Technical Paper Series 2, Centre for Migration Studies, University of Ghana, Legon.

Yaro, J. A., and D. Tsikata. 2015. "Recent Transnational Land Deals and the Local Agrarian Economy in Ghana." In *Africa's Land Rush: Rural Livelihoods and Agrarian Change*, edited by R. Hall, I. Scoones, and D. Tsikata, 46–64. Martlesham, UK: James Currey.

Yee, Amy. 2019. "Electronic Marvels Turn into Dangerous Trash in East Africa." *New York Times*, May 12.

Yeebo, Yepoka. 2014. "Inside a Massive Electronics Graveyard." *The Atlantic*, December 29.

Yu, Emily A., Matthew Akormedi, Emmanuel Asampong, Christian G. Meyer, and Julius N. Fobil. 2017. "Informal Processing of Electronic Waste at Agbogbloshie, Ghana: Workers' Knowledge About Associated Health Hazards and Alternative Livelihoods." *Global Health Promotion* 24 (4): 90–98.

Yusoff, Kathryn. 2018. *A Billion Black Anthropocenes or None*. Minneapolis: University of Minnesota Press.

Zahara, Alex. 2016. "Refusal as Research Method in Discard Studies." *Discard Studies* (blog), March 21. https://discardstudies.com/2016/03/21/refusal-as-research-method-in-discard-studies/.

Zeiderman, Austin. 2015. "Spaces of Uncertainty: Governing Urban Environmental Hazards." In *Modes of Uncertainty: Anthropological Cases*, edited by Limor Samimian-Darash and Paul Rabinow, 182–200. Chicago: University of Chicago Press.

Zimring, Carl A. 2015. *Clean and White: A History of Environmental Racism in the United States*. New York: New York University Press.

Zoellner, Tom. 2020. *Island on Fire: The Revolt That Ended Slavery in the British Empire*. Cambridge, MA: Harvard University Press.

Zuboff, Shoshonna. 2019. *The Age of Surveillance Capitalism: The Fight for a Human Nature at the New Fronteir of Power*. New York: Public Affairs.

Zuboff, Shoshanna. 2021. "The Coup We Are Not Talking About." *New York Times*, January 29.

Zuez, David. 2018. "Digital Afterlife: (Eco)Civilizational Politics of the Site and the Sight of E-Waste in China." *Anthropology Today* 34 (6): 11–15.

Zukin, Sharon. 2020. *The Innovation Complex: Cities, Tech, and the New Economy*. New York: Oxford University Press.

Index